OTHER TITLES OF INTEREST

BP74	Electronic Music Projects
BP173	Computer Music Projects
BP174	More Advanced Electronic Music Projects
BP182	MIDI Projects
BP185	Electronic Synthesiser Construction
BP246	Musical Applications of the Atari STs

MORE ADVANCED
MIDI PROJECTS

by

R. A. PENFOLD

BERNARD BABANI (publishing) LTD
THE GRAMPIANS
SHEPHERDS BUSH ROAD
LONDON W6 7NF
ENGLAND

PLEASE NOTE

Although every care has been taken with the production of this book to ensure that any projects, designs, modifications and/or programs etc. contained herewith, operate in a correct and safe manner and also that any components specified are normally available in Great Britain, the Publishers do not accept responsibility in any way for the failure, including fault in design, of any project, design, modification or program to work correctly or to cause damage to any other equipment that it may be connected to or used in conjunction with, or in respect of any other damage or injury that may be so caused, nor do the Publishers accept responsibility in any way for the failure to obtain specified components.

Notice is also given that if equipment that is still under warranty is modified in any way or used or connected with home-built equipment then that warranty may be void.

© 1989 BERNARD BABANI (publishing) LTD

First Published — July 1989

British Library Cataloguing in Publication Data
Penfold, R. A.
  More advanced midi projects.
  1. Music. Applications of computer systems
  I. Title
  780'.28'5416

ISBN 0 85934 192 5

Printed and Bound in Great Britain by Cox & Wyman Ltd, Reading

Contents

	Page
Chapter 1	
MIDI MESSAGES	1
Getting the Message	1
Modes	2
Mode 1	3
Mode 2	4
Mode 3	4
Mode 4	5
Multi-Mode	5
Transmission Modes	8
Note On/Off	8
Key Pressure	10
Controls	11
Mode Change	12
Program Change	14
Pitch Bend	15
System Messages	15
Song Position Pointer	15
Song Select/Tune Request	16
System Exclusive	16
System Real-time	18
Code Numbers	20
Channel Message	20
System Messages	21
The Hardware	23
Chapter 2	
SIMPLE MIDI ACCESSORIES	27
MIDI Indicator	27
THRU Box	32
The Circuit	34
MIDI Merge	36
The Circuit	41
Chapter 3	
MIDI PROJECTS	45
Micro or UART	45

Chapter 3 (continued) Page

- 6402 UART 46
- MIDI Code Generator 50
- Hex Codes 54
- Hex — Decimal — Binary Conversion Chart 55
- Output Circuit 55
- Power Source 57
- Construction 58
- Testing 63
- MIDI Pedal 64
- The Circuit 68
- Construction 72
- Setting Up 75
- MIDI Programmer 75
- System Operation 77
- The Circuit 79
- Construction and Use 86
- MIDI Processing 91
- System Operation 92
- The Circuit 94
- Construction 98
- In Use 101
- Simplification 101
- MIDI Analyser 102
- System Operation 102
- The Circuit 106
- Construction 110
- Finally 112
- Semiconductor Pinout Details 113

Preface

MIDI has not exactly been an overnight success, but after a slow start it has gained in popularity to the point where it now seems to dominate the world of electronic music. MIDI sockets are not only found on synthesisers – they are common on electronic pianos, electronic organs, and are even to be found on some guitars, audio mixers, effects units, computers, and portable keyboards. While in some spheres of the electronics industry a lack of true standardisation has led to promising products failing to gain acceptance by the consumers, MIDI has had the opposite effect on the sales of electronic music equipment. Compatibility between instruments etc. from different manufacturers is guaranteed provided they all have MIDI interfaces. Sales of electronic music equipment have never been better.

Although MIDI equipment tends to be considered by many as beyond the scope of the do-it-yourself enthusiast, this is not entirely true. It is probably not possible for the amateur builder to compete with electronic instrument manufacturers. Their use of specially produced integrated circuits and large volume production enables them to build instruments of a sophistication and low cost that the amateur could almost certainly never match. However, there are a number of useful MIDI accessories that are well within the capabilities of the average electronics enthusiast. These include such things as simple merge units and channelisers. Several such units are described in this book. These projects are generally more complicated than those featured in "MIDI Projects" (BP182), although a few very simple units are included. While most of these projects are not suitable for beginners, they should be well within the capabilities of someone who has a reasonable amount of experience at electronics construction. These circuits should also provide some useful electronic building blocks for use in readers' own designs.

R. A. Penfold

Chapter 1

MIDI MESSAGES

This book carries on where "MIDI Projects" (BP182) left off. In "MIDI Projects" a number of computer interfaces were described, together with some other reasonably simple MIDI projects. The projects described here are generally a little more advanced than those in book number BP182, and do not include any computer interfaces. The emphasis is more on units to extract the full potential from a MIDI system, whether or not it is computer based. The projects are "advanced" not only in that the circuits are generally more complex than those in BP182, but also in that they are mainly aimed at users of more sophisticated MIDI systems. With the current low prices of some MIDI expander units you no longer need to be a millionaire in order to put together an advanced music system!

The units described range from a simple MIDI merge unit through to more complex devices such as channelisers. I suppose that the projects fall into two main groups: those that are designed to overcome a deficiency in an item of equipment in the system, and those that are designed to enhance the performance of the system or to make it easier to use. Most modern items of electronic music equipment that implement MIDI have a comprehensive MIDI specification. In fact many modern instruments go somewhat beyond the MIDI standard, with "multi" mode and the like. Even with units of this type, there are still one or two simple gadgets that can enhance the performance of a system or its ease of use. Older MIDI equipment often has what is a considerably less than full implementation, and add-on units can be especially useful with equipment of this type.

Getting the Message

In BP182 the only MIDI messages that were considered were the basic note on and note off types. These are only the "tip of the iceberg" though, and there are a large number of different MIDI messages. Although in the past MIDI tended to be regarded as nothing more than an alternative to the old gate/CV method of interfacing, as many users now realise, its scope goes well beyond these confines. There is in fact no limit on the type of information that can be exchanged via MIDI. There are MIDI codes for specific types of message, and general codes that can be used to carry virtually any information you like. The only real limitation is the operating speed of MIDI, which means that the exchange of large amounts of information can take quite a long time.

In order to understand the function of some of the projects featured in this book you need to be familiar with many of these additional MIDI message types. In this first chapter a reasonably detailed description of all the MIDI messages will therefore be provided. It is well worth having a good knowledge of the MIDI messages as some of the less well known ones can be very useful. It is worth mentioning that it is a good idea to carefully study the MIDI specifications of any MIDI instruments or other MIDI equipment you own. Practically all equipment does not implement every MIDI feature, or only partially implements certain features. Studying the MIDI specifications for your equipment should make it clear what facilities are available to you and (more importantly) which ones are absent or not fully implemented. The MIDI implementation chart should also give details of any "extras" that go beyond the basic MIDI specification.

Modes

Before looking at the various types of MIDI message it would be as well to examine the subjects of channels and operating modes. Unless you understand the four MIDI modes and channelling, a lot of MIDI messages will be difficult or impossible to fully understand.

The concept of MIDI channels is quite easy to understand. At the beginning of each MIDI message there is some data which specifies the message type (note on, note off, etc.). Most messages also carry a channel number in this initial part of the message, so that they can be directed to one particular instrument in a system, or even to one voice of an instrument in the system. These are the MIDI "channel" messages. Some messages do not carry a channel number, and are directed at the entire system. Appropriately, these are called MIDI "system" messages.

MIDI channels are notional rather than real, since any device receiving a MIDI instruction can act on any channel number the message may contain, in whatever way the equipment designer chooses. This includes simply ignoring channel numbers! Remember that MIDI channels are simply numbers at the beginning of messages, and that they are not channels in the sense of separate connecting cables. It is for this reason that there are several MIDI modes, and these operating modes only differ in the way that MIDI channel numbers are treated. In all other respects they are the same.

With MIDI channel messages the most significant nibble (i.e. the four most significant bits) carry the message type code, while the least significant nibble contains the channel number. This gives sixteen channels from 0000 in binary (0 in decimal) to 1111 in binary (15 in decimal). Note that although the values used to select MIDI channels run from 0 to 15, the convention is to number MIDI channels from 1 to 16. Therefore, the value used in a MIDI message to select a channel is one less than the channel number (e.g. 10 is used in order to place a message on channel 11).

System messages have the appropriate code for the most significant nibble (11111 in binary — 240 in decimal). The least significant nibble is not needed for the channel number, leaving it free to identify different types of system message.

Mode 1

Mode 1 is alternatively known as "omni on/poly", and was originally called just "omni" mode (and often still is). This is the most simple mode, and the one to which many instruments default at switch-on. The "omni" part of the name means that channel numbers are ignored, and that an instrument in this mode will respond to any note on and note off messages that are received regardless of the channel number they contain. Exactly how received notes are assigned internally to an instrument's voices depends on the design of the instrument, and there is no standard for this. In most cases an instrument in mode 1 will respond to notes received via the MIDI input in exactly the same way as it would respond to the same sequence played on its keyboard.

This mode is intended as a basic mode which should enable any piece of MIDI equipment to function to some degree in conjunction with virtually any other piece of MIDI gear. It lacks versatility though, and is far from ideal for most sequencing. It is fine if you are using a single instrument which has all its voices producing the same sound, but it is not much use for anything else.

Mode 2
This has the alternative name of "omni on/mono". Like mode 1, channel numbers are ignored in this mode. The "mono" part of the name indicates that only monophonic operation is provided. This mode was presumably included in order to accommodate monophonic synthesisers, but few monophonic instruments equipped with MIDI interfaces have ever been produced. This mode is not normally included on polyphonic instruments, as there is no obvious advantage in downgrading them to monophonic types (which is effectively what would happen by switching to this mode)!

Mode 3
This is a powerful mode which does acknowledge the existence of MIDI channels. It has the alternative name of "omni off/poly". The "omni off" section of the name indicates that channel numbers are recognised, while the "poly" part indicates that polyphonic operation is possible. In other words, an instrument in mode 3 will only respond to notes on one channel, and the instrument will work with more than one note at a time switched on. MIDI does not set down any maximum or minimum number of notes that a mode 3 instrument must be able to handle at once. It is up to the user to ensure that an instrument used in this mode is not supplied with more notes than it can handle. With most instruments there is no major disaster if they should be fed with more notes than they can accommodate. Usually it simply results in existing notes being cut short so that new ones can be sounded (or with a lot of recent instruments any excess notes are ignored with existing ones sounding for their full duration). This is obviously something that should still be avoided if at all possible.

This mode has great potential for sequencing applications, since it is possible to have a number of instruments on separate MIDI channels providing different sounds. This gives you a sort of computer controlled orchestra, and tremendous potential to develop complex pieces of music. This mode was originally called "poly" mode incidentally, and is still occasionally referred to by this name.

Mode 4

This mode is widely regarded as the most powerful one for sequencing purposes, although I suppose that this is not strictly true. Its alternative name is "omni off/mono", but it is probably still better known by its original "mono" name. As the "omni off" part of the name implies, MIDI channel numbers are recognised in this mode. The "mono" part of the name is perhaps a little misleading in that it suggests that a mode 4 instrument can only provide monophonic operation. This is not true though, and operation is monophonic only in that each voice of the instrument operates monophonically. If an instrument has sixteen voices, then in mode 4 each voice is assigned to a different MIDI channel, and overall, sixteen note polyphonic operation is possible. This may not seem to be much better than the basic mode 1, but it gives tremendous scope when applied to a multi-timbral instrument. Each channel can then have a different sound, and a single instrument can provide a computer controlled orchestra.

This sort of system is actually less powerful than a mode 3 type having a number of instruments, because in mode 4 each channel only gives monophonic operation. It is popular with MIDI sequencer users because it gives extremely good results at an affordable price. A sixteen channel mode 4 instrument (or two eight channel mode 4 instruments) should cost substantially less than sixteen mode 3 instruments!

Multi Mode

This is an unofficial mode which is not mentioned in the MIDI specification. However, it seems to be a feature of many of the more recent MIDI instruments. It is only fair to point out that the exact implementation of this mode varies from one instrument to another, and it may be encountered under some other name. The basic idea of multi-mode is to provide polyphonic operation on several channels. It is rather like mode 4, but each channel is not restricted to monophonic operation. Another way of looking at it is to regard a multi-mode instrument as being like a number of mode 3 instruments set to operate on different channels. Because of this, the term "virtual instrument" is sometimes used to describe a channel of a multi-mode instrument.

The number of notes available on each channel varies from one instrument to another. So does the flexibility with which tone generators of an instrument are assigned to channels. With some instruments you have to specify a maximum number of notes for each channel, or select one of several preset allocations. With other instruments each channel can have as many notes at once as the instrument can provide. This is not to say that with, for example, a thirty-two note polyphonic instrument you could have thirty-two notes playing simultaneously on all sixteen MIDI channels. However, you could have thirty-two notes playing on each MIDI channel in turn, or eight notes playing on each of any four channels, or any combination which gives no more than thirty-two notes playing at once.

A good multi-mode instrument obviously provides great power for the sequencer user. I have heard demonstrations of multi-mode instruments which sound remarkably like a full orchestra playing! An apparent drawback of multi mode is that it is non-standard, and therefore would seem to potentially raise the likelihood of incompatibility problems. In practice this is really not a problem. As far as the device driving a multi-mode instrument is concerned it is dealing with several mode 3 instruments. From the MIDI point of view, a multi-mode instrument is just several mode 3 types in the same box.

An important point to realise is that you do not need to have all the instruments in a system working in the same mode. It is perfectly alright to have something along the lines of the system shown in Figure 1.1. Here a computer based sequencer is controlled two mode 4 instruments on channels from 1 to 14, and an eight channel polyphonic instrument in mode 3 on channel 15. These three instruments could not be

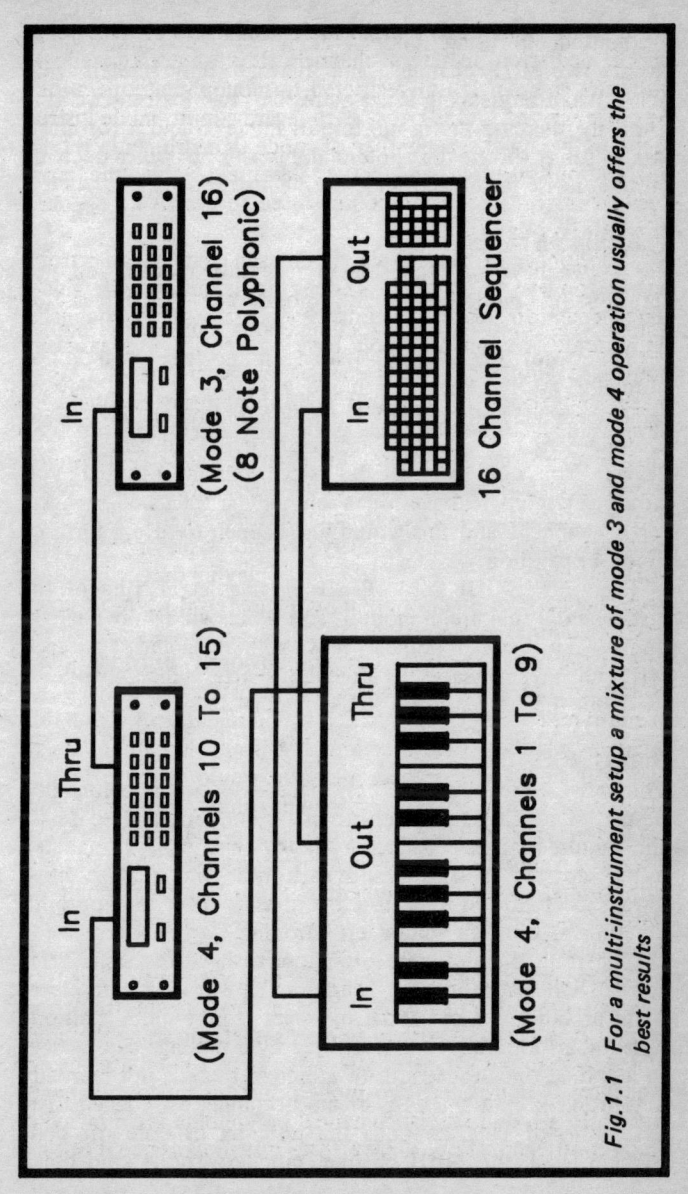

Fig.1.1 For a multi-instrument setup a mixture of mode 3 and mode 4 operation usually offers the best results

accommodated using mode 4 alone, as this would require twenty-two MIDI channels, and there are only sixteen available. When sequencing using more than two instruments it is normally the case that a mixture of modes 3 and 4 (or multi mode) offers the greatest potential, but this obviously depends on the precise facilities offered by the instruments.

Transmission Modes
So far we have only considered reception modes, and not transmission modes. Really, modes are a standard for handling received data, and talking about transmission modes is perhaps not totally valid. With a sending device you do not normally set an operating mode as such. If you have a sequencer giving monophonic operation on four channels, then I suppose that it could be accurately described as a mode 4 device. You would not set the device to mode 4 though, you would set it up to have four monophonic tracks on separate MIDI channels, and this would just happen to give a form of mode 4 operation.

Even if a MIDI controller is driving an instrument (or instruments) that are in mode 1, and which will ignore channel numbers, a channel number must still be included in each MIDI channel message. It does not matter which channel is used, but the convention is to use channel 1.

Note On/Off
For basic sequencing it is only necessary for the controlling device to be able to switch notes on and off, and to select the required notes. Switching notes on and off is handled using separate messages, and each note on message must always be followed by a note off message after the appropriate interval, or notes will be left "droning".

The note on and note off messages have the same basic format with each message being sent as three separate pieces of information. Most MIDI messages require more than one byte of information, but as a three byte message can be sent in less than one-thousandth of a second there is not normally any problem with the data stream becoming overloaded. The first byte of data contains the note on or note off code number, plus the MIDI channel number. The second block carries the note number, and MIDI supports a note range of 0 to 127. Each increment by one represents an increase in pitch by one semitone. A range of 128 semitones is a compass of well over ten octaves, which should be more than adequate. This is over three octaves more than the range covered by most pianos. In fact few MIDI equipped instruments actually accommodate the full note range. It is useful to bear in mind that many can handle a wider compass via their MIDI interface than they can using their keyboard. Incidentally, middle C is at a note value of 60.

The third byte in the message carries the velocity value for the note, which with most instruments controls the volume of the note. Most instruments are velocity sensitive these days, but there are still some that are not, and many early MIDI instruments did not implement this feature. This data must always be present though, so as to maintain full compatibility between items of MIDI equipment. A non-touch sensitive instrument simply ignores any velocity information it receives, and transmits an intermediate "dummy" value in any note on messages it transmits (usually a value of 64 is used). Once again, the range of values is from 0 to 127, representing minimum and maximum velocity respectively. MIDI data bytes always have the most significant bit set to 0, which makes them easily distinguished from message bytes where the most significant bit is always set to 1. Thus the range of data bytes is from 0 to 127, and not the 0 to 255 range one normally expects from eight bit codes.

Note off messages only differ from the note on type in that the message code in the first byte is different. It might seem at first that not all the information provided in note off messages is actually needed, but remember that there could be sixteen polyphonic instruments connected to a MIDI output. A note off message must make it clear which note on which channel of which instrument must be terminated. There is an alternative method of switching off notes, which is to use note on type having a velocity value of 0. I am not quite sure why this alternative method was deemed necessary, but some instruments do seem to use it (the SCI synthesisers in my original MIDI system certainly seemed to use this method for all note off operations, and apparently many others do so as well).

Key Pressure

At one time very few instruments responded to any form of key pressure (after-touch), but it now seems quite common for overall key pressure to be implemented. This is the most basic type of key pressure response, and it is a sort of average pressure value for however many keys are pressed. There are various ways in which this information can be used by an instrument, but it normally controls either the volume or the filtering after the initial attack and decay phases of the signal.

This type of message requires only two bytes of data to be sent, and the first of these contains the overall pressure code number together with the MIDI channel number for the message. The second block of data is the pressure value, which is from 0 (minimum pressure) to 127 (maximum pressure).

Polyphonic key pressure is much the same as the overall type, but individual MIDI messages are sent for each note. This is much more sophisticated, and offers excellent control, but it is relatively difficult to implement. It is a feature that, as yet, is far from common. The polyphonic key pressure message takes the same basic form as the overall type, but a third block of data is needed to identify the note to which the pressure value applies. This byte is placed between the code number and pressure value bytes.

An important point to keep in mind when using instruments that implement either form of key pressure is that holding down keys can result in a lot of MIDI data being generated. This applies more to the polyphonic type than those which have overall key pressure. With polyphonic key pressure, if you are holding down around five keys at a time, five sets of key pressure data will be transmitted. Key pressure is not something that is sent once per note, and there can be several sets of pressure data for each note. The implications of this for real-time sequencing are clear — you could easily end up with most of the available memory being taken up with pressure data rather than note on/off messages! This might not matter with short sequences if a lot of free memory is available. Otherwise, it might be necessary to disable the instrument's after-touch, or to use some method of filtering to remove the pressure data.

Key pressure is not something that is restricted to use with keyboard instruments. A lot of MIDI rack-mount modules will respond to after-touch, and there is no reason that a step-time sequencer should not be designed to implement key pressure. However, if you have equipment that can handle this type of thing, the memory problem described previously must be kept in mind.

Controls

MIDI includes a general purpose control message which can be used to control master volume, filter resonance, or anything a manufacturer cares to implement. The only MIDI control number which the MIDI specification allocates to a specific function is control number 1, which is the modulation wheel. It seems to be the convention for control 4 to be a foot pedal, control 7 to be the main volume, and for control number 64 to be the sustain pedal. However, these are only conventions, and not all equipment necessarily conforms to them.

Controller messages are standard three byte types, with the first byte carrying the controller message code and the channel number. The next chunk of data is the number of the control, and the last one is its new value (from 0 to 127). Control numbers from 0 to 63 are used for continuous controls (i.e. adjustable types like volume and tone controls) whereas control numbers from 64 to 95 are used for switches and only give a simple on/off action. For these switch type controls, only values of 0 (off) and 127 (on) are valid, and other values will be ignored.

The continuous controls are complicated slightly by a system of pairing, which has control numbers 0 to 31 paired with controls 32 to 63 respectively. The idea is to have the values sent to a pair of controls merged to give one large number. This permits much more accurate settings than are possible using one control in isolation. Whereas one control has 128 different settings from 0 to 127, a pair of controls gives 16384 settings from 0 to 16383. In practice the degree of control provided by pairs of control values is usually higher than is needed. Also, where a control is being continuously varied, such fine resolution requires vast amounts of data to be sent. It is unlikely that MIDI could send data fast enough to fully utilize such high resolution. In practice it seems to be quite common for the "fine tuning" controller not to be implemented. It is then only control numbers from 0 to 31 that are used, while those from 32 to 63 are ignored. As a point of interest, in several instruments I have used not even the full 0 to 127 range has actually been used, with some controls only having 64 or 32 different settings.

Mode Change

Control numbers from 96 to 127 are either not assigned to any purpose, or are used for things such as mode changing. The functions of the control numbers that have been assigned a task are listed below:—

Control Number	*Function*
121	Reset all controls
122	Local control on/off
123	All notes off
124	Omni mode off
125	Omni mode on
126	Mono mode on (poly mode off)
127	Poly mode on (mono mode off)

These are all switches where the values sent to them is either 0 (off) or 127 (on) or a dummy data number of 0 is used. The only exception is control number 126. Here the value sent specifies the number of voices to be used in mono mode (a value of 0 sets all available voices to mono mode).

Local control on/off is where the normal (built-in) method of controlling the instrument can be disabled. This usually means switching off the keyboard. One reason for doing this is merely to prevent accidental operation of an instrument while it is being sequenced from a computer. It also enables a keyboard instrument to effectively be used as a separate keyboard and sound generation module. The salient point here is that the keyboard will still transmit on the MIDI "OUT" socket, and the instrument will respond to data received on the MIDI "IN" socket. You can even feed the MIDI output though some form of processor (which could be a computer running a suitable program or a stand alone unit), and then

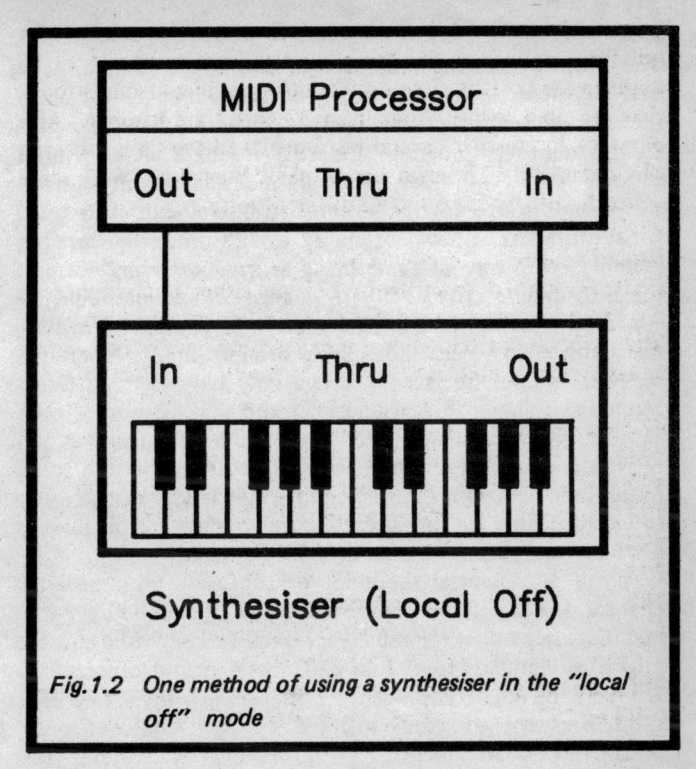

Fig.1.2 One method of using a synthesiser in the "local off" mode

feed the processed signal back into the instrument, as shown in Figure 1.2.

The all notes off message is not intended as the normal way of switching off notes. It seems to be intended more as a means of switching off any "droning" notes in the event of some form of malfunction. Incidentally, changes in MIDI mode also switch off any notes that are switched on at the time.

MIDI does not have specific messages to select mode 1, mode 2, etc., but instead the right combination of omni on/off, mono, and poly have to be selected. This should all be quite straightforward if you consider modes in terms of their current names rather than their numbers (e.g. omni on/poly instead of mode 1).

Program Change

The program change message uses two bytes of data. The first of these is the appropriate message code and channel number, while the second is the new program for that channel. In this context "program" generally means a set of control settings for a synthesiser, so we are really talking in terms of a change in sound for the voice of an instrument. It is usual for synthesisers to have around 32 to 128 programs stored in memory, with any of these being assignable to any voice of the instrument. The ability to change sounds mid-sequence via MIDI is more than a little useful. If you have a synthesiser with eight voices and capable of multi-timbral operation in mode 4, on the face of it you only have eight different sounds available. If the sequencer and the synthesiser both support program changes, then you could have as many as 128 different sounds available. You would still be limited to no more than eight notes at once, but would have a vast range of sounds available for use in each piece of music. Not quite as good as having a bank of synthesisers, but nearly!

There is potential for a lot of confusion with program changes as there is no standard method of numbering. One manufacturer might use numbers from 0 to 127, while another might use numbers from 1 to 128. Some manufacturers have a totally different approach. For instance, my Casio CZ1 synthesiser has programs selected by two banks of pushbuttons which are labelled "1" to "8" and "A" to "H". This gives sixty-four programs from "A-0" to "H-8". The manual for an instrument should make it quite clear if there is a discrepancy between the numbers of programs, and the values used in program messages to select them. Often there is a chart showing program change values and which program each one selects.

It is worth noting that program change messages are not only recognised by synthesisers and other instruments. Devices such as MIDI controlled mixers and effects units often make use of them as well. The usual way in which this works is that sets of control adjustments are assigned to program numbers so that they can be called up as and when required using the appropriate program change instructions. Program changes are an important part of much MIDI sequencing.

Pitch Bend

Pitch bending could be accomplished using an ordinary MIDI controller, but it has been assigned its own MIDI message type. This message consists of three bytes, with the first one containing the pitch bend message code and the channel number. The other two bytes of data contain the pitch bend value, with the two numbers having to be combined into one large pitch bend value at the receiving device. The MIDI specification does not lay down rules stipulating exactly how much given changes in pitch bend value actually affect the pitch of an instrument. Pitch bend information recorded from one instrument might not produce exactly the same degree of bend if it is played back to a different instrument. Some instruments now permit some control over the degree of pitch bend produced by a given change in the pitch bend value.

System Messages

System messages have a variety of functions, but they are mainly concerned with timing, and the synchronisation of sequencers. This almost invariably means keeping the built-in sequencer of a drum machine properly synchronised with the main sequencer which controls the rest of the system. There is another important category of system messages in the form of "system exclusive" types, which seem to be playing an increasingly prominent role in the world of MIDI.

Song Position Pointer

A MIDI sequencer which is capable of using MIDI synchronisation signals can keep track of the number of beats that have elapsed since the start of a sequence (or "song"). The maximum number of beats that can be handled is 16384, and each beat is equal to 1/16th note. The idea of this is to enable a sequencer to randomly access any part of a song. Perhaps more accurately, the idea is to enable two sequencers operating in tandem to be set to exactly the same point in a song, and any desired point in a song. Without this random access feature they could only be kept in synchronisation by always starting them both from the beginning of a sequence, or by using a lot of trial and error.

The song position pointer uses two bytes of data, but these two numbers are combined to give one large number of 0 to 16383. Note that some sequencers take a significant time to adjust to a song pointer instruction, and that they must be given time to respond to one of these messages before they are restarted. Also note that this message only moves the sequencers to a certain point in a song, it does not set the sequencers in motion.

Song Select/Tune Request

Although these messages have names that suggest a similar function, they are actually quite different. The song select message is used to select the desired sequence from a sequencer that can store more than one song, and which supports this feature. As with virtually every type of MIDI message, do not assume that your equipment actually implements this feature. Always check the MIDI specifications very carefully to find out what features are supported and which are ignored. This message uses one data byte, and this is the song number. Song numbers are from 0 to 127 in terms of the actual number sent in a song select message. However, this is another example of the identification numbers used by manufacturers not necessarily being the same as the actual values used in the MIDI message, and not necessarily being the same from one manufacturer to another.

In the tune request message it is "tune" in the sense of tuning an instrument. This is for use with instruments which have an automatic tuning facility. The message only consists of the message code with no data being sent. No timing information is sent with this message, and all it is really doing is telling instruments to tune themselves against their internal tuning references. Presumably any instruments that implement this feature would then accurately tune themselves to the usual pitch of $A = 440$ Hz, and would all be accurately in tune with each other. This is not a feature that seems to be much used these days, and few instruments seem to support this MIDI message.

System Exclusive

Most MIDI messages are universal, and can be implemented by

any equipment manufacturer. This is an important aspect of MIDI, as one of the prime reasons for its introduction was, as far as possible, to eliminate incompatibility between devices from different manufacturers. On the other hand, MIDI needed to be flexible enough to permit future expansion and developments. Manufacturers needed to be able to do their "own thing", and implement any novel ideas that they might develop. Without this built-in flexibility it was unlikely that MIDI would have been adopted by all the main electronic music equipment producers.

Much of MIDI's flexibility lies in the system exclusive message. This consists of the message code followed by the manufacturer's identification code. The idea of this code is that it enables equipment to filter out and ignore system exclusive messages that do not have the correct manufacturer's identification number. This is an important feature, because the data that follows the identification code can be anything the equipment manufacturer desires. The data here will either be meaningless to the wrong piece of equipment or worse still, it could have totally the wrong effect and render a piece of equipment temporarily useless. The system is only usable with this method of filtering included. There is no fixed number of data bytes in a system exclusive message, and there can be as much data here as the application requires. The end of a system exclusive message is marked by a special (single byte) MIDI message.

System exclusive messages are used for such things as program dumps, sample dumps, or any non-standard feature that an equipment producer wishes to implement via MIDI. A number of recent instruments seem to use system exclusive messages to provide MIDI control of their sound generator circuits, rather than making these adjustable via standard MIDI controller messages. This is perhaps an unfortunate trend, as units (or a computer like the ST plus some software) that are intended for general MIDI programming via controller messages are not usable with these system exclusive oriented instruments. They can only be programmed by way of a matching programmer unit, or custom software.

On the other hand, manufacturers who use system exclusive messages are supposed to publish details of the method of coding used, and to allow anyone to freely use this coding. Once system exclusive details have been published, no changes should be made to the specification (except perhaps, to extend it rather than modify any existing details). It is quite in order for a "third party" company to produce programming software or hardware that accesses instruments via system exclusive messages, and many programs of this type are available.

There have been moves towards standardising some system exclusive messages. As far as I know, the only standard system exclusive message at present is the MIDI sample dump standard. This uses the sample dump standard identification code where the manufacturer's code number would normally be, and it then has quite a complex method of sending samples. This complexity is inevitable, as this standard is designed to accommodate a wide range of instruments at different levels of sophistication. It is also designed to leave sufficient "headroom" for future developments. It has facilities for error checking, and has two way communications so that a receiving device can temporarily halt the flow of data if it is becoming overloaded. However, it is only fair to point out that not all samplers use the sample dump standard at the moment, and it might never be adopted as the only sample dump standard.

Incidentally, this type of system exclusive message is sometimes known by the rather contradictory name of "system exclusive common" message.

System Real-time

The system real-time messages are the ones which provide synchronisation between two sequencers, and are analogous to the clock pulses used to synchronise drum machines in the pre-MIDI era (and probably still much used today). The MIDI system is substantially different to the old system though. In particular, the clock signal is not just a regular series of electronic pulses. It is a regular series of MIDI clock messages, and it is sent continuously, not just while a sequence is in progress. Because of this a number of other messages are needed in order to make the system workable. It should be pointed out here that in the original MIDI specification synchronisation was handled in a slightly different manner.

This original method is now completely obsolete though, and is certainly not to be found on any currently produced equipment.

With a continuous clock signal it is obviously necessary to have stop and start messages so that sequences can be started and halted as required. Actually, there are two types of start message, "start" and "continue". They differ in that a "start" message results in the sequence starting from the beginning, whereas continue causes it to start from wherever it left off (i.e. the current position of the song pointer). If a song pointer instruction is used to move to the middle of a sequence, it is "continue" and not "start" that should be used to restart the sequence.

The system real-time messages include a "reset" instruction, which simply takes the equipment back to its initial state (i.e. the state in which it would be if you were to switch it off and then turn it on again). This is not implemented on all instruments, and would not be worthwhile with many disc based instruments such as samplers, which can not produce any sound in their switch-on state (they must first have data loaded from disc).

Another little used facility is active sensing. The idea here is that the MIDI controller sends out an active sensing message at reasonably frequent intervals (not more than 0.3 seconds between each one), and the controlled devices then check that they are receiving these messages at suitable intervals. If a gap of more than 0.3 seconds should elapse without an active sensing message being received, all notes are terminated. This is quite a good idea as it avoids having an instrument stuck with notes activated or in some other "hung-up" state if a connecting cable becomes damaged, or something of this sort should occur.

This facility seems to be little used in practice though. It has the disadvantage of increasing the amount of MIDI data that is transmitted, which increases the risk of MIDI "choke". This would not seem to be a major problem though, as it only needs three to four messages per second to be transmitted. Perhaps equipment manufacturers feel that the processing power of their instruments and controllers could be put to better use in other MIDI departments, or perhaps they feel that this is an unnecessary complication that few people will want to bother with. Anyway, you are unlikely to encounter much equipment that actually implements this feature.

All these system real-time messages are single byte types, and none of them contain any data. Being forms of system message they do not carry a MIDI channel number either.

Code Numbers

This is a full list of the standard MIDI messages, and if you are intending to use anything more than very simple MIDI systems you need to be familiar with many of these. In particular, if you are going to use the projects described in the next two chapters of this book you should study the details of the channel messages. The next section of this book consists of several tables which provide details of the MIDI code numbers. These are of little interest to many MIDI users, but if you intend to write MIDI software, design your own hardware, or do trouble-shooting when a system does not behave as anticipated, details of the MIDI codes are essential. It is probably not worthwhile trying to commit the MIDI code numbers to memory, but this section of the book should be useful for reference purposes.

Channel Messages

Table 1

Header	Function	Data
1000 (128)	Note Off	Note Value/Velocity Value
1001 (144)	Note On	Note Value/Velocity Value
1010 (160)	Poly Key Pressure	Note Value/Pressure Value
1011 (176)	Control Change	Control Number/Value
1100 (192)	Program Change	New Program Number
1101 (208)	Overall Pressure	Pressure Value
1110 (224)	Pitch Wheel	l.s.b./m.s.b.

Table 2

Control No.	Function	Data
122	Local Control	0 = off, 127 = on
123	All Notes Off	0
124	Omni Mode Off	0
125	Omni Mode On	0
126	Mono Mode On	Number of channels (0 = All Channels Set To Mono Mode)
127	Poly Mode On	0

System Messages

These all have 1111 as the most significant nibble in the header byte. No channel numbers are used, as these messages are sent to the whole system. This leaves the least significant nibble free to indicate the type of system message. Table 3 gives a full list of these messages, but note that some of the sixteen available codes are as yet undefined. Many of them do not require data bytes, and are just single byte messages.

Table 3

Nibble Code	Function	Data
0000 (0)	System Exclusive	ID/As Required
0001 (1)	Undefined	
0010 (2)	Song Position Pointer	l.s.b./m.s.b.
0011 (3)	Song Select	Song Number
0100 (4)	Undefined	
0101 (5)	Undefined	
0110 (6)	Tune Request	None
0111 (7)	End System Exclusive	None
1000 (8)	Clock Signal	None
1001 (9)	Undefined	
1010 (10)	Start	None

continued overleaf

Table 3 (continued)

Nibble Code	Function	Data
1011 (11)	Continue	None
1100 (12)	Stop	None
1101 (13)	Undefined	
1110 (14)	Active Sensing	None
1111 (15)	System Reset	None

The values shown in brackets are the decimal equivalents for the binary nibbles. These must be boosted by 240 to give the total decimal value for each header byte (e.g. the value sent for a clock signal is 240 + 8 = 248). The system exclusive message is followed by a data byte which gives the manufacturer's identification code, and then as many data bytes as required follow on from this. The "end system exclusive" message marks the end of a system exclusive message. Table 4 provides a list of manufacturer's identification numbers. The sample dump standard is a "system exclusive common" message, which can be used by any MIDI equipment producer.

Table 4

Manufacturer	Number (decimal)	Manufacturer	Number (decimal)
SC1	1	Bon Tempi	32
Big Briar	2	SIEL	33
Octave	3	Kawai	64
Moog	4	Roland	65
Passport Designs	5	Korg	66
Lexicon	6	Yamaha	67
Ensonique	15	Casio	68
Oberheim	16	Sample Dump Standard	126

The Hardware

MIDI is a form of asynchronous serial interface, and in this respect it is very much like ordinary computer RS232C and RS423 interfaces. The standard MIDI "baud" rate is 31250 baud, or 31.25 kilobaud if you prefer. This is not a standard RS232C baud rate, and might seem to be an unusual choice. Originally the baud rate was 19200 baud, which is the highest standard baud rate for RS232C interfaces. However, this was deemed to be too slow, and in the final MIDI specification it was increased to 31250 baud. This is convenient from the hardware point of view, as it is well within the capabilities of most serial interface chips. Also, 31250 multiplied by 32 equals 1000000, and this fact enables the baud rate of MIDI interfaces to be controlled using "off the shelf" crystals intended for communications applications and microprocessor circuits.

RS232C and RS423 interfaces use different voltages to represent logic 0 and logic 1 levels, but MIDI is different in that it uses a 5 milliamp current loop. In other words, the current is switched on to indicate one logic level, and switched off to represent the other logic state. This is done due to the use of opto-isolators at each input, which keep items of equipment in the system electrically isolated from one another. This eliminates the risk of damage occurring when two or more items of equipment are connected together, due to their chassis being at different voltages. It also helps to reduce the risk of "hum" loops being produced when a number of instruments and other equipment are connected together. Finally, it also helps to avoid having electrical noise coupled from a computer or other micro controller circuit into the audio stages of an instrument. If there is one thing computers do better than space invaders or number crunching it is generating electrical noise! MIDI port connection details are shown in Figure 1.3.

There is a slight problem in using opto-isolation with a system that has a relatively high baud rate of 31250 baud. Opto-isolators are inherently slow devices, and this rate of switching is well beyond the capabilities of the popular low cost types such as the TIL111. At least, it is beyond their capabilities unless they are augmented with a switching

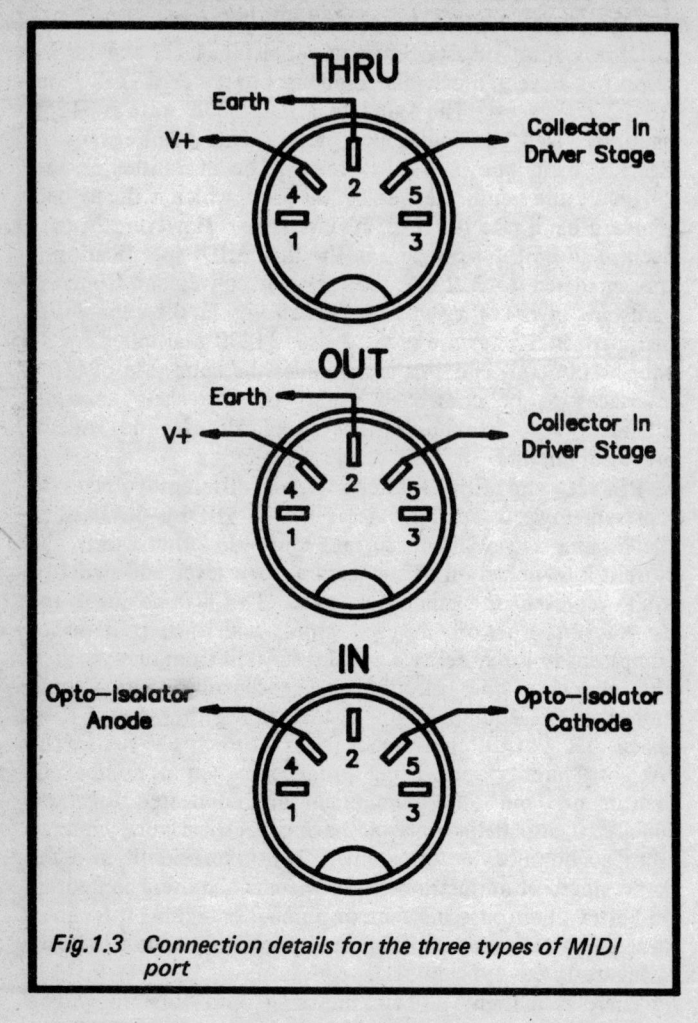

Fig.1.3 Connection details for the three types of MIDI port

transistor at the output to boost their performance. I have found that the arrangement shown in Figure 1.4 usually gives excellent results.

Of course, there are various improved opto-isolators which offer increased efficiency and switching speed. Of these, the

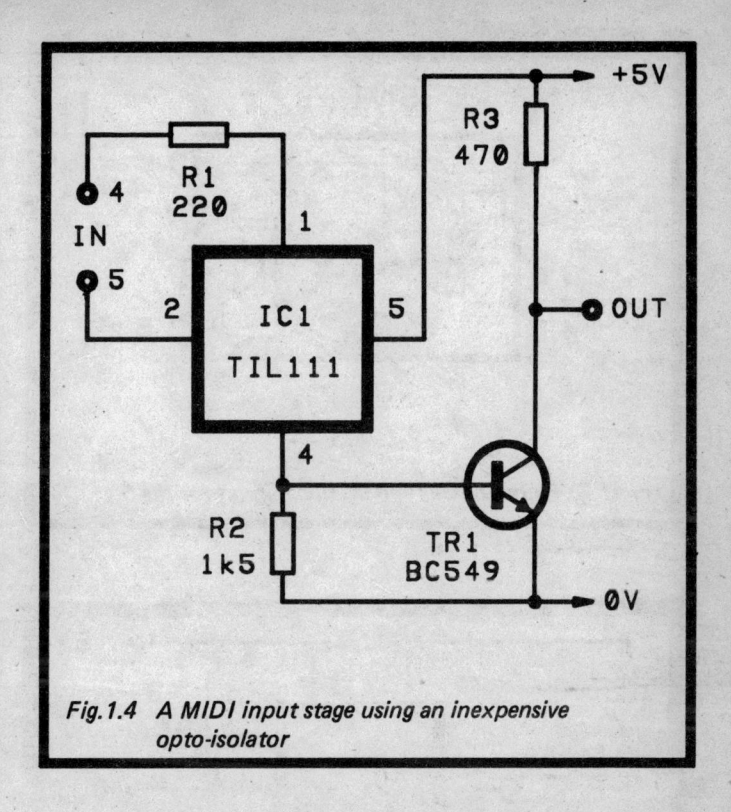

Fig.1.4 A MIDI input stage using an inexpensive opto-isolator

type which seems to offer the best results in a MIDI context are the ones which have an infra-red l.e.d. driving a photodiode, which in turn drives a two stage amplifier circuit. This arrangement is shown in Figure 1.5, which shows the internal circuit and pinout details for the 6N139. Figure 1.6 shows how this device can be used as a MIDI input stage. Opto-isolators of this type are relatively expensive, but they provide very good reliability. They are actually capable of operating at speeds of up to something like ten times the MIDI baud rate!

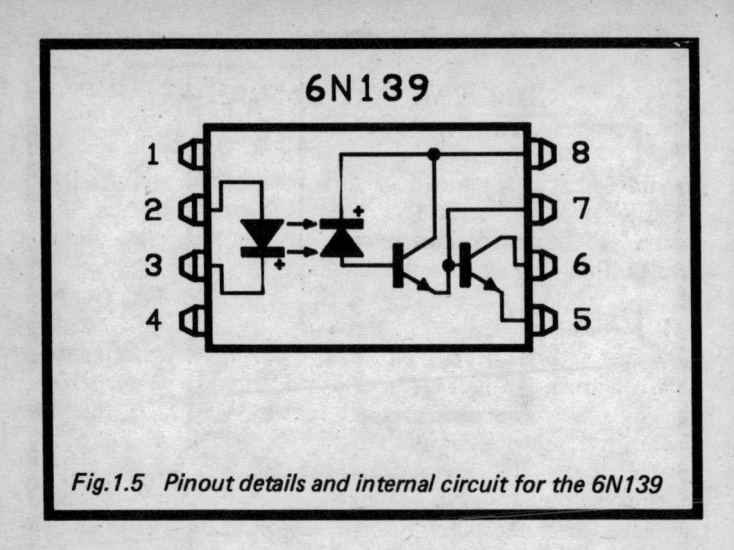

Fig.1.5 Pinout details and internal circuit for the 6N139

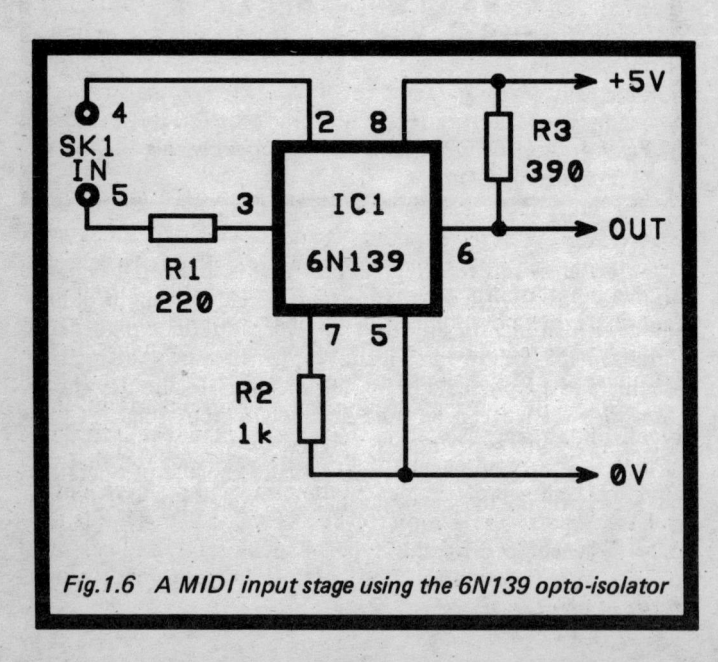

Fig.1.6 A MIDI input stage using the 6N139 opto-isolator

Chapter 2

SIMPLE MIDI ACCESSORIES

In this short chapter two or three simple but useful MIDI accessories are described. These are not really in the "advanced" category, but it was felt that they were useful gadgets that were well worth inclusion in this book.

MIDI Indicator

The first of these projects is simply a unit that connects into a MIDI cable and indicates via a l.e.d. whether or not MIDI messages are being sent over the cable. Many MIDI instruments and other MIDI units have a built-in l.e.d. indicator of this type, but it is by no means present on all MIDI equipment. Also, most built-in indicators are only activated when a MIDI message to which the unit must respond is received. In other words, if an instrument is set to operate on (say) channel 3, its MIDI indicator light will not respond to messages on anything other than channel 3. This can be advantageous in certain circumstances. On the other hand, it can sometimes lead to uncertainty as to whether the system is not set up correctly, or there is a hardware fault such as a broken cable.

The full circuit diagram for the MIDI indicator appears in Figure 2.1. Strictly speaking, a unit of this type does not need to have an opto-isolator at the input. The opto-isolator at the input of the unit fed from the indicator will ensure that there are no problems with "hum" loops etc. Being battery powered, there is little opportunity for this unit to introduce any problems of this type.

Despite this, and the added expense of opto-isolation, this type of input stage is used in the circuit. One reason for this is that it is a requirement of the MIDI specification that all equipment, however simple or complex, should have opto-isolated inputs. At a more practical level, MIDI outputs are only designed to drive this type of input stage, and can not be absolutely guaranteed to function properly with other types of input stage.

Fig.2.1 The MIDI indicator circuit

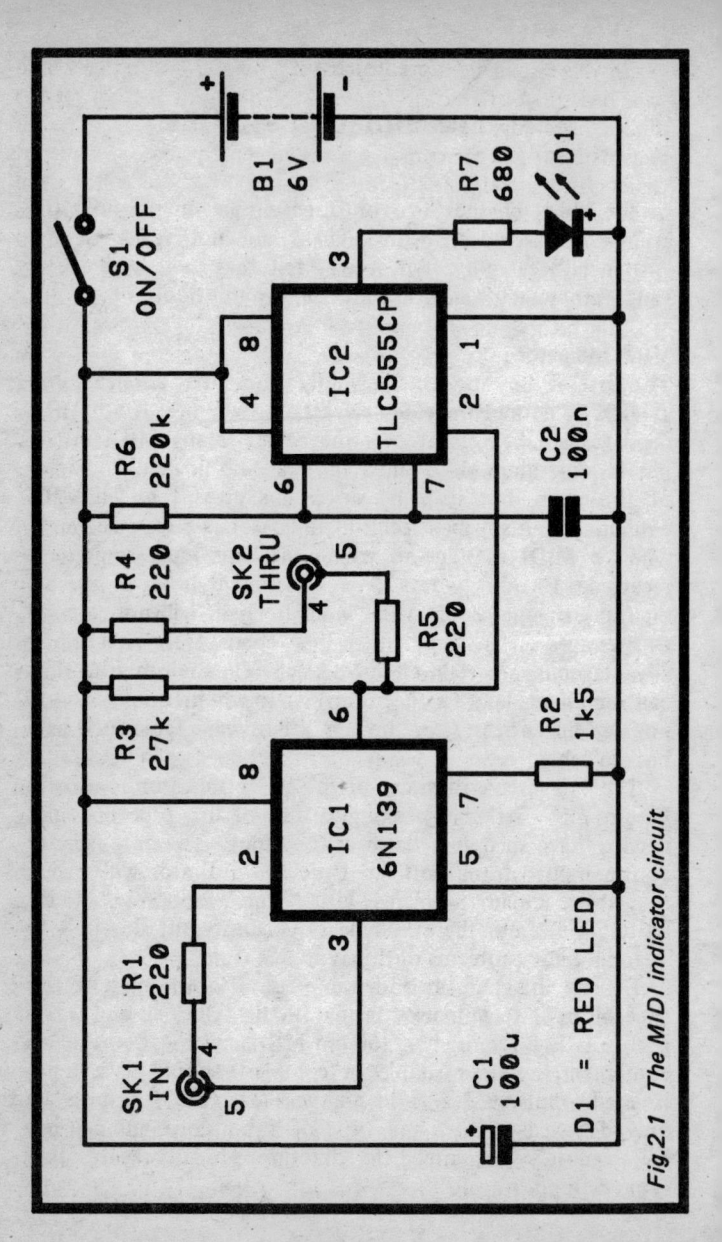

In this circuit the opto-isolator is a 6N139 — a device which was described briefly in Chapter 1. The input section of this circuit is similar to the input stage based on the 6N139 that was described in Chapter 1. However, in this case the output must drive a MIDI input, rather than a UART or other serial interface device. The output transistor of the 6N139 is therefore used as an open collector driver stage, in standard MIDI fashion. R1 provides current limiting at the input, while R4 and R5 provide the same function at the output.

The l.e.d. indicator could be connected in the output circuit of IC1, since this device is quite capable of driving an indicator l.e.d. at good current and driving a MIDI input. However, with only infrequent MIDI messages, especially ones which contained code numbers that caused the output transistor of IC1 to be switched off during most bits, the flashing of the l.e.d. could be so brief that it would be barely visible. Much better results are obtained using a pulse stretcher circuit to ensure that each flash of the l.e.d. is long enough to make it clearly visible.

In this circuit the pulse stretcher is a monostable multivibrator based on a 555 timer integrated circuit. The circuit should work perfectly well using the standard 555, but I would recommend the use of a low power version such as the TLC555CP or L555P. This reduces the current consumption of the circuit by about 5 milliamps and gives better battery life.

The monostable is a standard 555 type, and it is triggered when pin 2 of IC2 is taken below one-third of the supply voltage. This occurs when a MIDI message is received and the output transistor of IC1 switches on. R3 ensures that there is always a load resistor for IC1's output transistor, and that the unit will operate properly even if the THRU socket is left unconnected. R6 and C2 set the output pulse duration at roughly 250ms, which is long enough to ensure that a clearly visible flash is produced by D1. The l.e.d. current is only about 6 milliamps. Use a high brightness l.e.d. for D1 if the unit is likely to be used in high ambient light levels.

Current consumption from the 6 volt supply is largely dependent on how much data is passed through the unit. Under standby conditions the current consumption is about 200 microamps, but this figure will vary somewhat depending on which low power version of the 555 is used in the IC2 position. Even with a constant stream of data the current consumption is unlikely to be much more than about 10 milliamps. A six volt battery comprised of four HP7 size cells in a plastic holder should give at least a few hundred hours of operation. Incidentally, connection to this type of battery holder is usually via a standard PP3 style battery connector.

Construction of the unit should pose few difficulties. Although most low power 555 devices seem to be based on CMOS technology, most do not require anti-static handling precautions. They have very effective built-in static protection circuits. Neither IC1 nor IC2 are particularly cheap though, and I would strongly urge the use of holders for both components. Provided SK1 and SK2 are both 5 way (180 degree) DIN sockets connected in the appropriate manner, the unit will function properly with standard MIDI leads.

If you are making up your own leads, any reasonably good quality twin screened cable will normally suffice. MIDI is only guaranteed to operate at ranges of up to 15 metres, but in order to obtain something approaching this range it might be necessary to resort to a high quality cable. Most people only need MIDI cables about 2 metres or so in length, and virtually any cable will then suffice. The correct method of connection is to use one of the inner conductors to connect pin 4 on one plug to pin 4 of the other plug. The other inner conductor is used to connect the two pin 5s together. The screen connects the two pin 2s, while pins 1 and 3 are left unconnected. The correct method of wiring up a MIDI lead is shown in Figure 2.2.

Pin 2 of SK2 should be connected to the unit's negative supply rail. In practice, this might not be very effective in reducing the radiation of radio frequency interference from the lead. Better results might be obtained by interconnecting pin 2 of SK1 and pin 2 of SK2. Note though, that this method of interconnection should only be used on a simple, battery powered, in-line unit such as this one. With more major items of equipment pin 2 of the IN socket should always be left unconnected.

Fig.2.2 Connection details for a standard MIDI lead

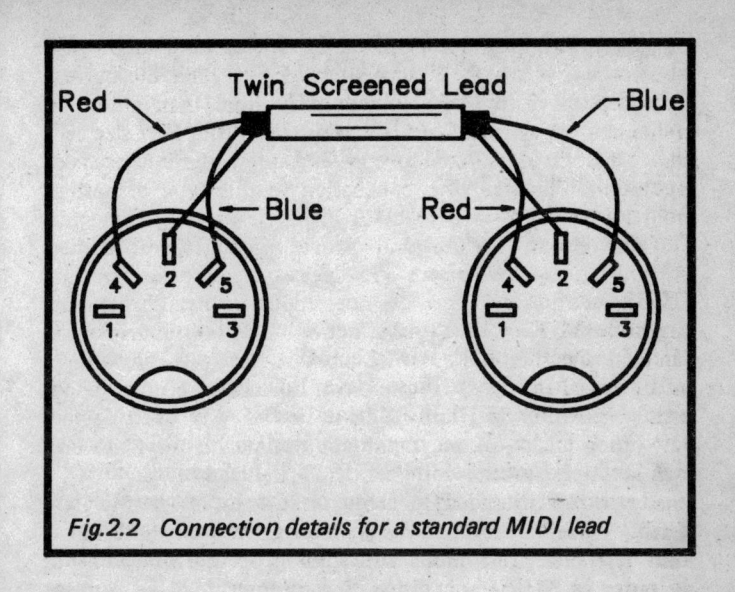

Components for MIDI Indicator (Fig.2.1)

Resistors (all 0.25 watt 5% carbon film)
R1	220
R2	1k5
R3	27k
R4	220
R5	220
R6	220k
R7	680

Capacitors
C1	100µ 10V elect
C2	100n polyester

Semiconductors
IC1	6N139 opto-isolator
IC2	TLC555CP or similar
D1	Red panel LED

Miscellaneous

B1	6 volt (4 x HP7 in plastic holder)
S1	SPST sub-min toggle
SK1	5 way (180 degree) DIN socket
SK2	5 way (180 degree) DIN socket

8 pin d.i.l. holders (2 off)
Battery connector
Circuit board, case, wire, etc.

THRU Box

A simple THRU box project was described in the original "MIDI Projects" book. It was actually called an "expander" in book BP182, but these days this term seems to be mainly used for MIDI instruments that lack a keyboard or any other means of playing them manually. These instruments can be in the form of 19 inch rack-mount units or smaller boxes intended to stand on the top of an electric organ. "THRU box" seems to be the accepted term for a unit that takes one input and splits it to provide multiple outputs, or THRU sockets as they become if MIDI terminology is strictly applied.

The point of a THRU box is that it avoids the so-called MIDI delays. The standard method of connecting several MIDI units to a controller is to adopt the "chain" method of connection. Here the THRU socket of the first unit is connected to the IN socket of the second unit, the THRU socket of the second unit connects to the IN socket of the third, and so on. The MIDI specification does not specify a maximum number of units that can be "chained" together in this way. In practice the signal tends to be degraded slightly as it makes each journey from an IN socket to a THRU type. With a number of units connected together in this way there is a real danger of the signal being corrupted before it reaches the final instrument in the setup.

Although this problem is popularly known as "MIDI delay", it seems unlikely that the problem has anything to do with delaying of the signal. The delay through even ten or twenty units is likely to be just a fraction of a millisecond. It is probably more the cumulative effect of frequency response restrictions through each piece of equipment, finally distorting

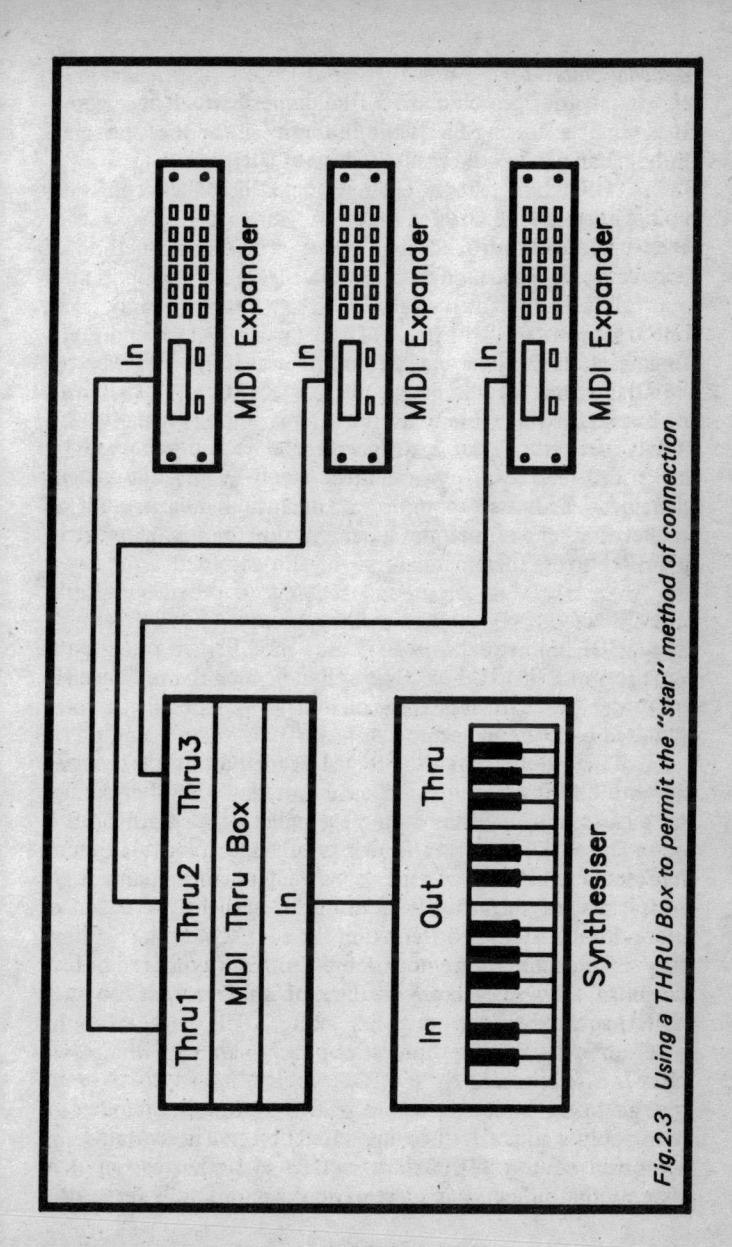

Fig.2.3 Using a THRU Box to permit the "star" method of connection

the waveform to the point where it is no longer readable by the serial interface chip of the equipment which receives it. It is really a slowing up of the rise and fall times of the signal rather than what could really be termed a delay.

A THRU box takes the output signal from the controller and splits it several ways so that it provides a THRU socket for each piece of controlled equipment. The "star" method of connection is then used, as shown in Figure 2.3. Each item of controlled equipment then receives a signal that has only taken one trip from an "IN" to a "THRU", and which should therefore have undergone no significant waveform degradation.

Apart from increased reliability, a THRU box is essential in a system where there are several instruments which lack a THRU socket. There is no problem if only one instrument lacks this facility — you simply use it at the end of the "chain". With two or more instruments that lack a THRU socket the "chain" method of connection is unusable. Use of a THRU box is then mandatory.

The Circuit

The MIDI indicator project is easily modified to provide the function of a THRU box while still indicating the presence (or otherwise) of an input signal. The additional circuitry required is shown in Figure 2.4.

This additional circuitry is just more output sockets and current limiting resistors added in parallel with the existing output circuit. This is quite acceptable since the 6N139 is quite capable of driving four sets of output sockets (which represents an output current of only about 20 milliamps). In fact it would probably be possible to have half a dozen or more THRU sockets driven from the 6N139, although I have to point out that I have not tried the unit with more than four outputs. Obviously there is a lack of isolation between each THRU socket, but this does not matter. The pieces of equipment driven from the unit should be isolated via their own opto-isolated inputs.

The notes on constructing and using the MIDI indicator unit apply equally well to the THRU box. The current consumption of the THRU box will be somewhat higher than that of the indicator unit as more outputs are being driven.

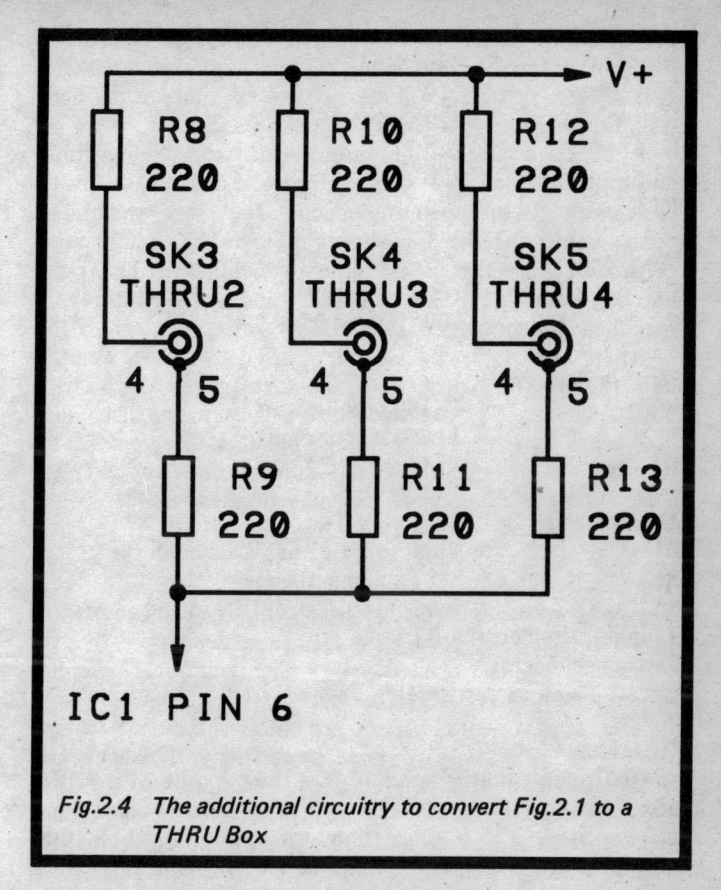

Fig.2.4 The additional circuitry to convert Fig.2.1 to a THRU Box

Components for THRU Box (Figs 2.1 & 2.4)

Resistors (all 0.25 watt 5% carbon film)
R1	220
R2	1k5
R3	27k
R4	220
R5	220
R6	220k
R7	680

Resistors (continued)
R8   220
R9   220
R10   220
R11   220
R12   220
R13   220

Capacitors
C1   100µ 10V elect
C2   100n polyester

Semiconductors
IC1   6N139 opto-isolator
IC2   TLC555CP or similar
D1   Red panel LED

Miscellaneous
B1   6 volt (4 x HP7 in plastic holder)
S1   SPST sub-min toggle
SK1 to SK5   5 way (180 degree) DIN socket (5 off)
8 pin d.i.l. holders (2 off)
Battery connector
Circuit board, case, wire, etc.

MIDI Merge

A MIDI merge unit is, more or less, the opposite of a THRU box. In other words, it takes two or more input signals and merges them into a single output signal. Remember that normally only one controlling device is permissible in a MIDI system, and that simply wiring two outputs together is a very risky business which is unlikely to give satisfactory results. Whether or not a merge unit is required depends on whether your system will ever need to have more than one controlling device. Probably the most common usage of a MIDI merge unit would be when a system contains both a sequencer of some kind and a keyboard. Sometimes you might want to control the system from the sequencer, while at other times you might prefer to play it "live". This can involve a lot of plugging-in and unplugging of MIDI leads.

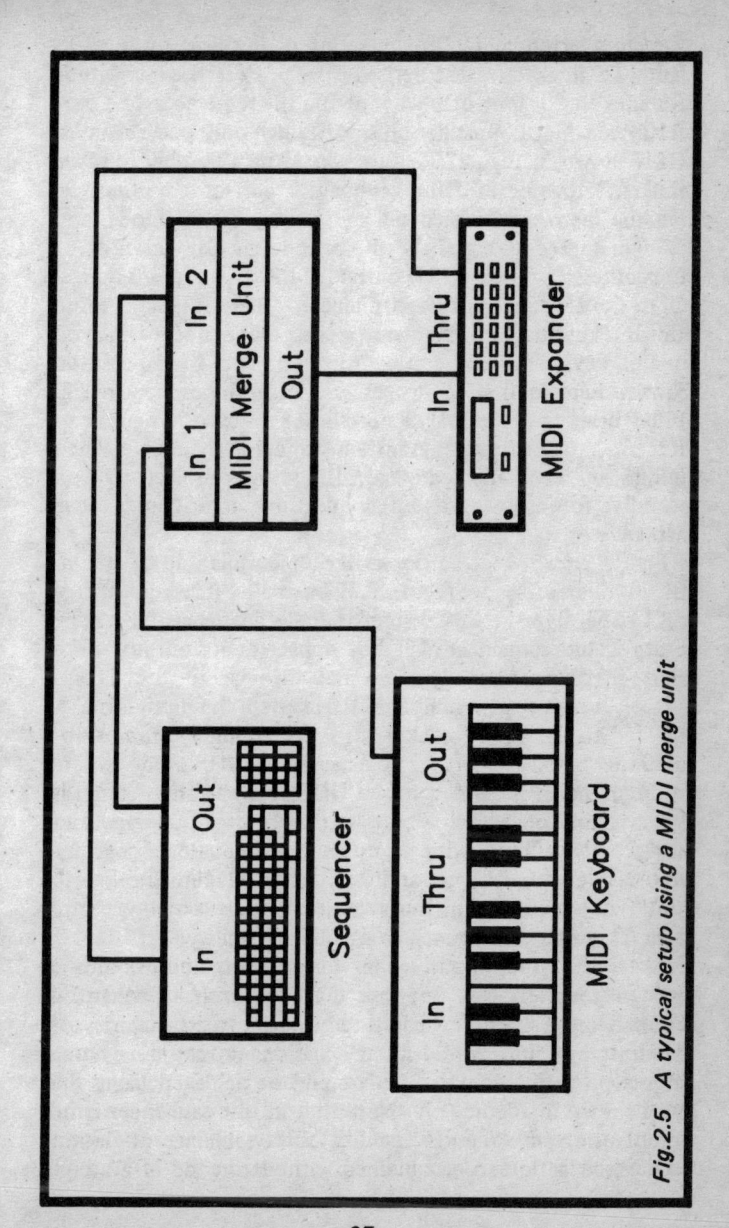

Fig.2.5 A typical setup using a MIDI merge unit

Using a MIDI merge unit a set-up of the type outlined in Figure 2.5 might give the desired effect. Here the merge unit combines the output of keyboard and the sequencer, and uses it to drive a MIDI expander unit. Although only one expander unit is shown in Figure 2.5, there could be a number of these "chained" together. If the keyboard is part of a synthesiser, then the instrument could be set to the local off mode, and the expander could then be the sound generator circuits of the synthesiser. The THRU output of the expander is connected to the IN socket of the sequencer. This provides a route from the keyboard to the sequencer so that sequences played on the keyboard can be recorded. There is a potential problem here in that sequences sent from the sequencer will be fed back to its input. This will not necessarily cause any problems, but in some circumstances it certainly could. When building up any fairly complex MIDI setup you need to look carefully for any unwanted connections that could cause problems.

Figure 2.6 shows what looks like a plausible way of obtaining a similar setup to Figure 2.5, but without using a merge unit. The flaw in this system is that the signal fed to the input of the sequencer will not appear at its output. The input signal is normally only fed through to the THRU socket. At least, in a standard MIDI system the input signal is only fed to the THRU socket. It is increasingly common for computer based sequencers to have a "THRU" mode, where the input signal is echoed to the OUT socket. This feature is also available on a few synthesisers. Obviously a merge unit should not be used with equipment that could obtain the same result without one, and it is worth reading the "small print" in equipment manuals to see if there are any useful extra features available such as a "THRU" facility.

There are other situations in which a merge unit could be used to good effect. Suppose that you wish to control a system from a computer based sequencer for the majority of the time, but that you wish to occasionally feed the system from a programmer unit. This could be achieved using the setup shown in Figure 2.7. Note that in this case there is no way of using any "THRU" facility of the sequencer to permit the desired action to be obtained without the aid of a merge

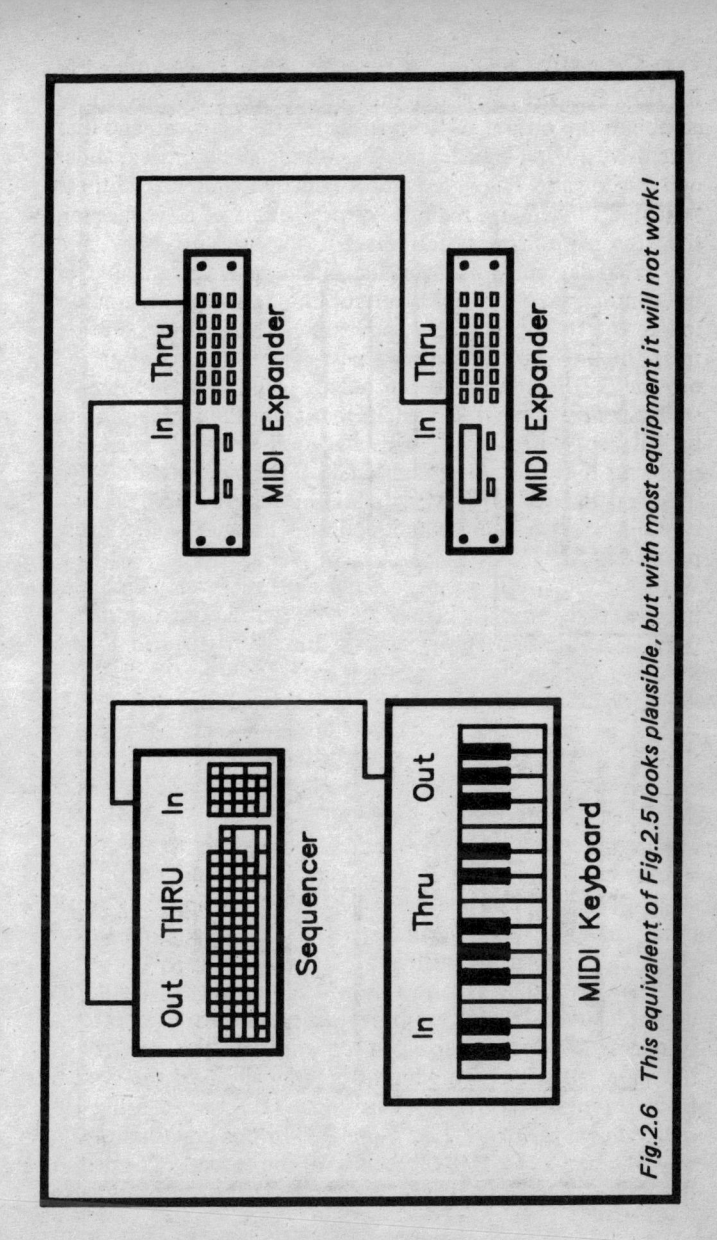

Fig.2.6 This equivalent of Fig.2.5 looks plausible, but with most equipment it will not work!

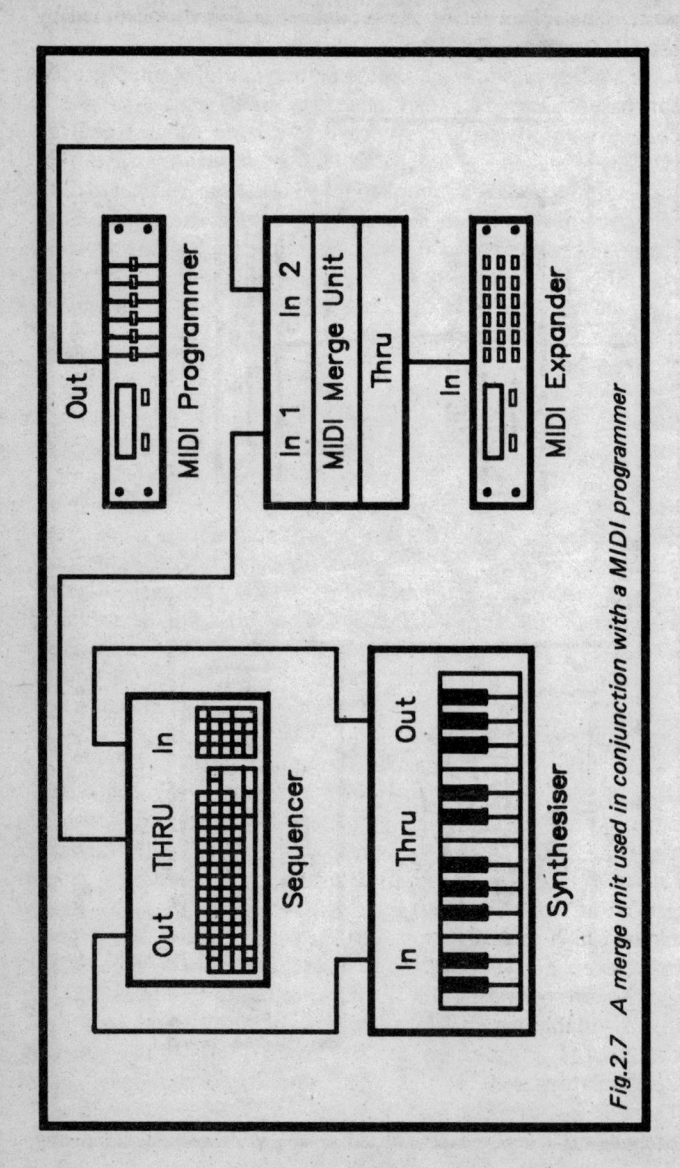

Fig.2.7 A merge unit used in conjunction with a MIDI programmer

unit. The IN socket of the sequencer is already occupied by the OUT signal of the keyboard instrument.

A sophisticated merge unit is a quite complex microprocessor based unit which can mix two input signals to give a coherent output signal. At least, it can do so provided the two input signals are not so heavily laden with messages that the output becomes "choked". This type of merge unit requires a buffer (i.e. a block of memory) that can be used to store messages received at one input while messages received at the other input are sent through to the output. In this way the unit can cope with messages sometimes being received on both inputs simultaneously.

The Circuit

A simple unit of the type described here simply mixes the two inputs, so that a signal on either input is coupled through to the output. However, if signals are received on both inputs at once, the output signal is a mixture of the two input signals. This is almost certain to give an output signal that is completely "scrambled" and unusable. This might seem to be a bit useless, but the salient point is that in many instances where a system has two controllers, only one of these will be used at a time. You can use a switch to select the desired controller, but a merge unit is more convenient in that, in effect, it automatically switches over to whichever of the controllers is producing an output signal.

The full circuit diagram of the unit appears in Figure 2.8. This consists of what is really just three opto-isolator circuits having open collector output stages that are connected in parallel. Thus, any one of the opto-isolators being activated results in an output current flowing, and there is a signal path from each input to the single output. Three 6N139 or similar high specification opto-isolators would be rather expensive. For economy the unit has therefore been based on inexpensive opto-isolators with a discrete output transistor to give a suitably high efficiency and switching speed. Although the TIL111 is specified for IC1 to IC3, similar inexpensive opto-isolators such as the 4N27 should work equally well. There is just a slight risk that one of the opto-isolators will not give sufficiently high efficiency or switching speed, causing

Fig.2.8 The MIDI merge unit circuit diagram

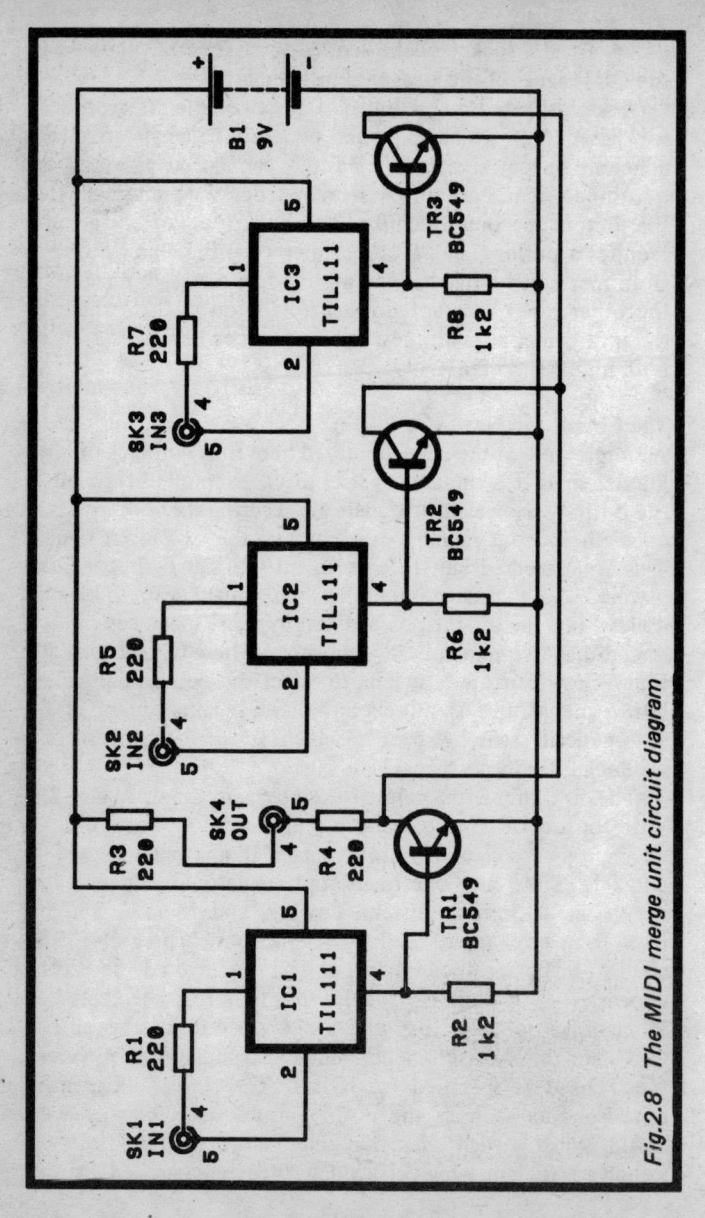

one of the inputs to operate unreliably. However, even if an extra TIL111 is obtained in order to guard against this eventuality, the unit can still be built quite cheaply.

The circuit is not fitted with an on/off switch as the current consumption under quiescent conditions is negligible. Obviously a SPST on/off switch could be added in the positive battery lead if desired. The current consumption of the circuit is largely dependent on the amount of data passed by the unit, but it should never average much more than about 5 milliamps.

Construction of the unit should be quite straightforward. You may not deem it worthwhile fitting the opto-isolators in sockets as they are inexpensive devices. I prefer to use holders for all d.i.l. components regardless of their cost, but in this case there is a slight difficulty in that the necessary 6 pin d.i.l. holders are not widely available. You might be able to track down suitable holders, but if not it is quite easy to trim an 8 pin d.i.l. holder down to size.

Although the unit is shown here as having three inputs, it can easily be cut back to two inputs if preferred, simply omit IC3, TR3, SK3, R7, and R8. Similarly, more inputs can be added if desired, and in theory any number of opto-isolator stages can be wired in parallel, giving any desired number of inputs.

Components for MIDI Merge (Fig.2.8)

Resistors (all 0.25 watt 5% carbon film)
R1	220
R2	1k2
R3	220
R4	220
R5	220
R6	1k2
R7	220
R8	1k2

Semiconductors
IC1	TIL111 or similar
IC2	TIL111 or similar
IC3	TIL111 or similar

Semiconductors (continued)
TR1   BC549
TR2   BC549
TR3   BC549

Miscellaneous
B1   9 volt (PP3 size)
SK1 to SK4   5 way (180 degree) DIN socket (4 off)
6 way d.i.l. holder (3 off, see text)
Battery connector
Circuit board, case, wire, solder, etc.

Chapter 3

MIDI PROJECTS

The projects in this chapter are concerned with the generation, decoding, and modification of MIDI signals. The topics covered include such things as a MIDI program change pedal and a channeliser. By modern standards these projects are not terribly complex, but they are certainly not suitable for complete beginners at electronic project construction. However, they should be within the capabilities of anyone who has had a moderate amount of experience at building electronic projects.

Micro or UART?

Something that these projects have in common is that they are based on the industry standard 6402 UART (universal asynchronous receiver/transmitter). In fact they will also work using the AY-3-1015D which is generally a little cheaper. The AY-3-1015D is not an exact equivalent for the 6402, but the differences are so minor as to be of no significance in the current applications.

The 6402 is a serial interface chip which provides both serial to parallel and parallel to serial conversion. Whereas many serial interface devices are only intended for operation in microprocessor based circuits, UARTs are general purpose devices that will operate in both micro and non-micro based circuits. They have tristate outputs, and inputs that are used for programming the word format etc. These can be used with the busses of a microprocessor, but they operate just as well driving l.e.d. indicators and being fed from programming switches. This second option is not applicable to most serial interface chips which rely totally on a microprocessor plus suitable software to set them up and to regulate the flow of data.

Of course, projects of the types featured in this chapter could be based on a simple microprocessor based circuit. One circuit plus suitable software routines could probably provide all the functions afforded by these projects! The problem with a microprocessor based circuit for these types of application is that it is simple only in microprocessor terms. It would be quite a complex and expensive gadget in normal project terms. It would require the constructor to have access to an EPROM programmer, or the purchase of comparatively expensive custom programmed EPROMS, as these represent the only practical method of program storage for units of this general type. Circuits based on a UART plus some ordinary logic integrated circuits were felt to be a more practical proposition for home constructor projects. None of the circuits described in this book are microprocessor based.

6402 UART

Understanding the way in which these projects function is very much easier if you are reasonably familiar with the 6402 UART. Pinout details for this device are shown in Figure 3.1, and an explanation of the function of each pin is given in the next section of this chapter.

Pin 1, V+
This is the positive supply terminal, and should be fed with a nominal 5 volt supply (4 volts minimum — 7 volts absolute maximum).

Pin 2, NC
There is no internal connection to this pin.

Pin 3, GND
The ground, or negative (0 volt) supply terminal in other words.

Pin 4, RRD
Receiver Register Disable. A high level on this input takes the 8 bit output of the receiver section to the high impedance state. In a non-microprocessor based unit this facility is not normally needed, and this pin would be tied to ground.

Pin 5 to Pin 12, RBR0 to RBR7
These are the parallel outputs of the receiver section of the unit. As soon as a byte has been received and decoded it is placed on these outputs.

```
        1           40
V+    ──┤ ○       ├── TRC
NC    ──┤         ├── EPE
GND   ──┤         ├── CLS1
RRD   ──┤         ├── CLS2
RBR7  ──┤         ├── SBS
RBR6  ──┤         ├── PI
RBR5  ──┤         ├── CRL
RBR4  ──┤         ├── TBR7
RBR3  ──┤         ├── TBR6
RBR2  ──┤         ├── TBR5
RBR1  ──┤         ├── TBR4
RBR0  ──┤         ├── TBR3
PE    ──┤         ├── TBR2
FE    ──┤         ├── TBR1
OE    ──┤         ├── TBR0
SFD   ──┤         ├── TRO
RRC   ──┤         ├── TRE
DRR   ──┤         ├── TBRL
DR    ──┤         ├── TBRE
RRI   ──┤         ├── MR
        20          21
```

Fig.3.1 Pinout details for the 6402 UART

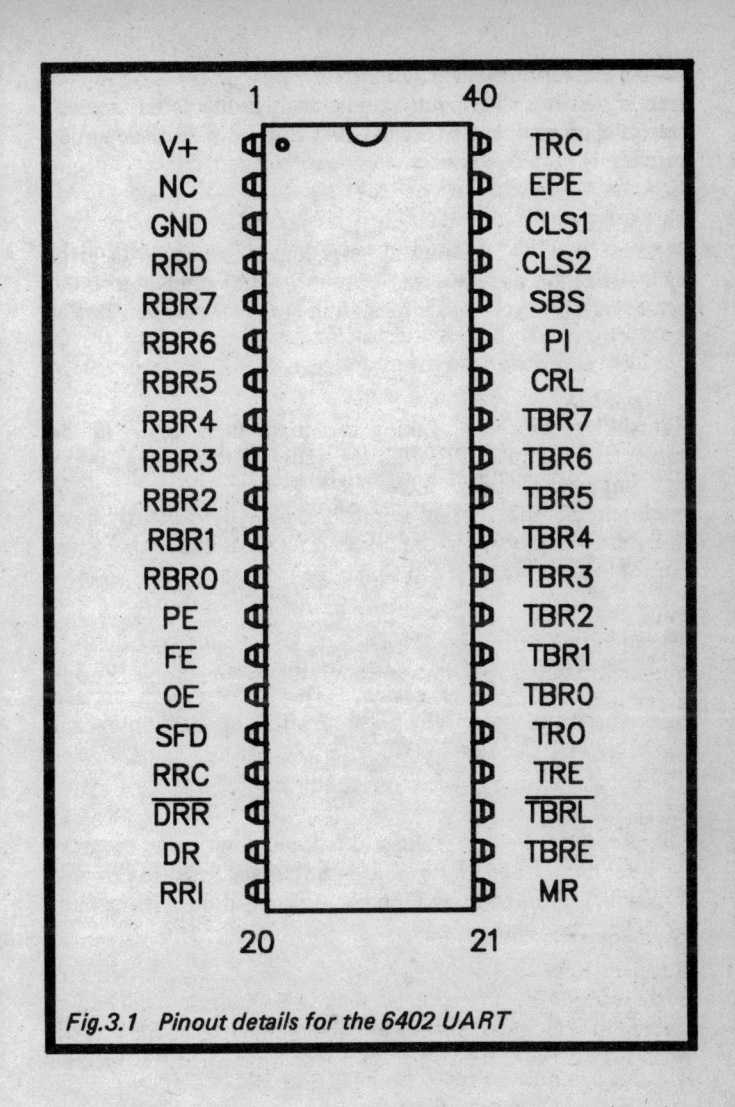

Pin 13, PE
Parity Error. This output goes high if the serial decoder circuit detects a parity error.

Pin 14, FE

Framing Error. This output goes high if the serial decoder detects a framing error (i.e. the first stop bit was at the wrong logic level).

Pin 15, OE

Overrun Error. This output goes high if an overrun error occurs (i.e. the data received output was not reset before the last received byte was transferred to the receiver buffer register).

Pin 16, SFD

Status Flags Disable. Taking this input high results in the status flag outputs (PE, FE, OE, DR, TBRE) going to the high impedance state. Like RRD, in a non-microprocessor based circuit this pin is normally taken permanently low. The status flag outputs can be used to drive indicator l.e.d.s, control logic circuits, or can simply be ignored.

Pin 17, RRC

Receiver Register Clock. This is the clock input for the receiver section of the device. The clock rate is sixteen times the baud rate, which for a MIDI application means 500 kHz (31250 x 16 = 500000).

Pin 18, DRR

Data Received Reset. Taking this input low resets the data received flag. There is no built-in automatic resetting of this flag, which is not really a practical proposition with a non-micro based system anyway.

Pin 19, DR

Data Received. This output goes high when a complete byte has been received and transferred to the receiver buffer register. This flag can only be reset using DRR or MR.

Pin 20, RR1

Receiver Register Input. This is the input to which the serial input signal is fed.

Pin 21, MR
Master Reset. Taking this input high clears the status flags, but not the receiver buffer register. This input must be fed with a positive pulse at switch-on in order to initialise the device (which becomes fully operational within 18 clock cycles of MR going low).

Pin 22, TBRE
Transmitter Buffer Register Empty. This output goes high when the transmitter buffer register has transferred its data to the transmitter register. In other words, a high level on this terminal indicates that the transmitter section of the device is ready to receive the next byte of data.

Pin 23, TRBL
Transmitter Buffer Register Load. A low level on this input results in the data on the parallel input of the transmitter section being transferred to the transmitter buffer register. As this input goes high again the data is transferred to the transmitter register, provided the latter is not currently sending a byte of data. If it is, then the transfer is automatically delayed until transmission of the current byte has been completed. Virtually all serial interface chips use a similar method of buffering at the transmitter's input.

Pin 24, TRE
Transmitter Register Empty. This output goes high when a byte of data has been fully transmitted, including the transmission of stop and any parity bit.

Pin 25, TRO
Transmitter Register Output. The serial output signal is produced on this pin.

Pin 26 to Pin 33, TBR0 to TBR7
Transmitter Buffer Register 0 to Transmitter Buffer Register 7. Parallel data for the transmitter section of the UART is coupled to these eight pins.

Pin 34, CRL
Control Register Load. This input is taken high in order to load the control register with the data fed to these five inputs (pins 35 to 39). In a non-micro based circuit this input can be taken high permanently.

Pin 35, PI
Parity Inhibit. A high level on this input switches off parity checking and generation. MIDI does not use parity checking, and this input is therefore taken high in the circuits featured in this book.

Pin 36, SBS
Stop Bit Select. This input is taken low for 1 stop bit operation, or high for 1.5/2 stop bit operation. It must be taken low in order to give the single stop bit required for MIDI use.

Pins 37 and 38, CLS1 – CLS2
Character Length Select 1 and 2. The binary code set on these inputs selects a character length of 5, 6, 7, or 8 bits. MIDI operation requires 8 data bits, and this mode is obtained with both CLS1 and CLS2 taken high.

Pin 39, EPE
Even Parity Enable. Taking this input low sets odd parity – setting it high selects even parity. This assumes that parity has been enabled using pin 35. In a MIDI application parity is disabled, and the logic level on pin 39 is therefore unimportant.

Pin 40, TRC
Transmitter Register Clock. This is the clock input for the transmitter section of the device. Like the receiver clock input, the clock frequency must be 16 times the required baud rate.

MIDI Code Generator
It can often be useful to have a unit that can generate MIDI codes. This can be necessary for testing purposes, or you might find that certain functions can only be obtained by sending an instrument the proper MIDI codes. For example,

Fig.3.2 The main MIDI code generator circuit

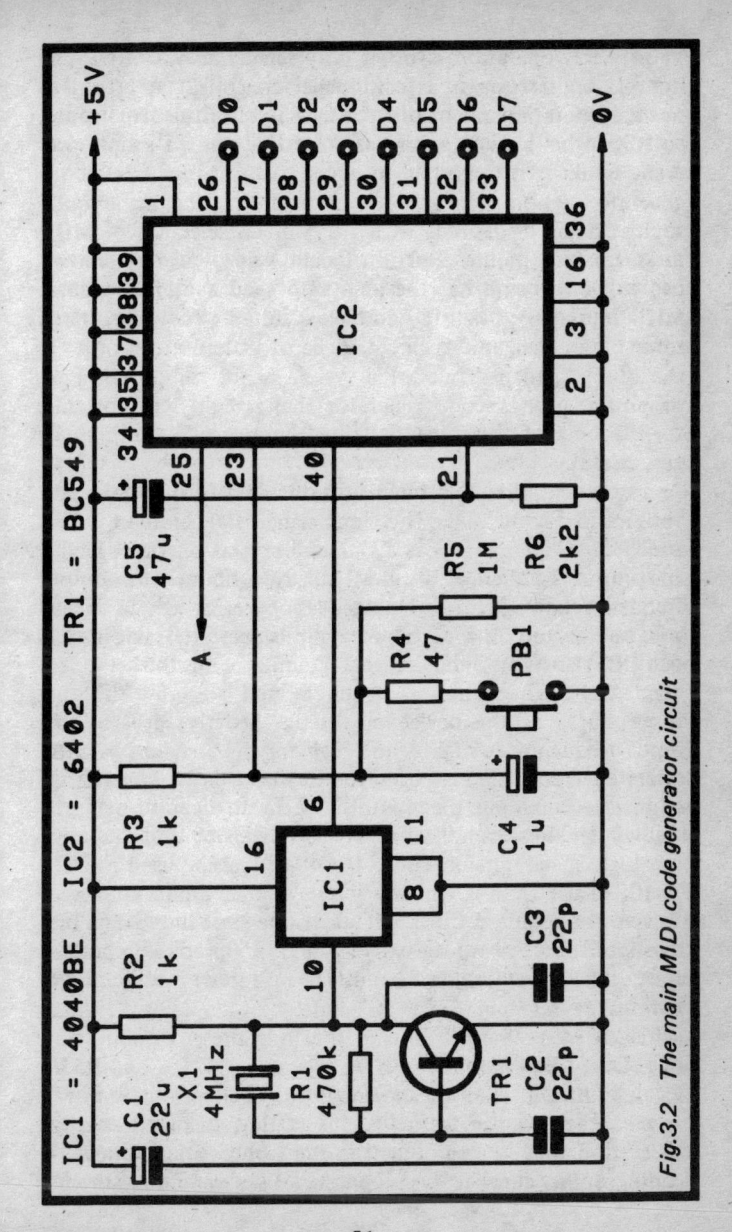

"local off" operation is often only selectable via MIDI, and not via an instrument's front panel controls. A return to normal operation might only be available via the same route.

If you have a computer based MIDI system, a simple program is all that is needed in order to provide a MIDI code generator action. In fact the program can be made quite sophisticated if desired, with (say) menu selection of MIDI messages and mouse control of data bytes. Some commercial MIDI software has the ability to send a useful range of MIDI message types. If you have a suitable computer based setup, then this unit is likely to be of little use to you. On the other hand, if your MIDI system is not computer based, or you require a code generator that is light and portable, then a unit of the type described here should prove to be very useful.

The main circuit diagram for the MIDI code generator appears in Figure 3.2. The clock signal is generated by TR1 and IC1. TR1 operates as a standard crystal oscillator having an output frequency of 4 MHz. No trimmer for output frequency adjustment is included as any error in this respect will be far too low to be of significance. IC1 is a CMOS 4040BE 12-stage binary ripple counter. In this circuit a divide by eight action is required, and so only the first three stages of the device are used. Strictly speaking, the input frequency is slightly too high for the standard 4040BE operated from a 5 volt supply. However, in practice CMOS counters (including the 4040BE) seem to operate perfectly well under these conditions. To be sure of correct operation the high speed version (the 74HC4040) can be used for IC1.

IC2 is the UART, and this has its control inputs connected to give the required word format of one start bit, 8 data bits, one stop bit, and no parity. C5 and R6 produce a positive reset pulse at switch-on. In order to transmit a byte of data from the unit, pin 23 of IC2 must be pulsed low. This is achieved by pressing PB1, and the byte present on the data bus (D0 to D7) is sent as PB1 is released. C4, R4, and R5 are a "debouncing" circuit, and should be adequate to deal with any contact bounce from PB1. I would recommend the use of a good quality component in the PB1 position though, as some of the cheaper types produce very generous amounts

Fig.3.3 The Hex code generator circuit

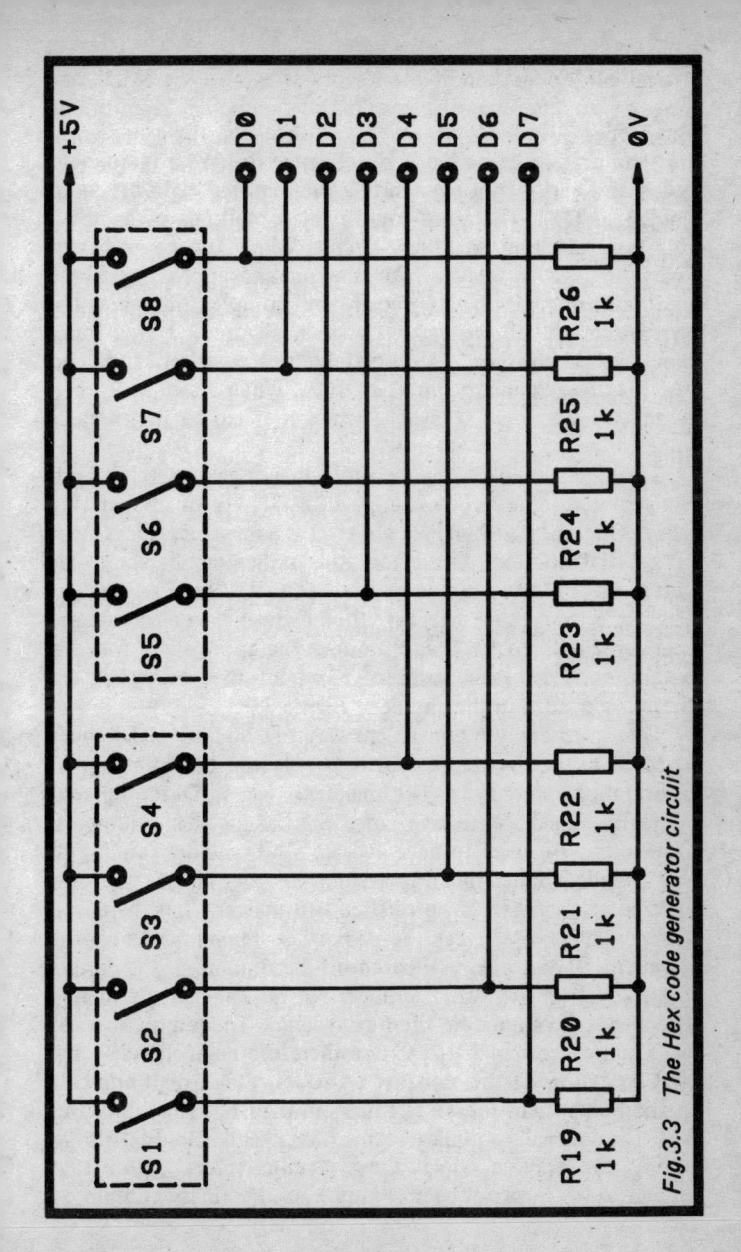

of contact bounce.

Hex Codes

For the unit to be usable it is necessary to have a reasonably quick and easy way of generating the required eight bit codes on D0 to D7. There are sophisticated methods of handling this type of thing, including circuits which use keypads and l.e.d. or l.c.d. displays. For the present application these methods are somewhat too costly and complex though, and a switch circuit of the type shown in Figure 3.3 is a more practical proposition. This merely uses a switch and load resistor at each binary input of IC2. When a switch is open it generates a logic 0 level — when it is closed it generates logic 1.

S1 to S8 could simply be individual toggle switches with the required binary code being set on them prior to operating PB1. Operating at binary code level is sometimes convenient in a MIDI context, but often it is rather cumbersome. A better option is to use two hexadecimal ("hex") switches. These are four pole rotary or "thumbwheel" switches that are numbered "0" to "F", and produce the appropriate four bit binary code at each position. Two of these switches are therefore needed in the circuit of Figure 3.3.

This is generally a convenient way of handling MIDI codes as MIDI header bytes are split into two four bit "nibbles" of information anyway. The manuals for MIDI equipment generally make extensive use of hexadecimal numbers. Obviously you need to have at least a basic understanding of the hexadecimal numbering system in order to use the unit efficiently, but hex is not difficult to master. It is based on the number 16, whereas the decimal system is based on the number 10, and binary is based on the number 2. There are not enough single digit numbers in the decimal numbering system to accommodate the hex system. The ten numbers of the decimal system (0 to 9) are therefore augmented by the first six letters of the alphabet (A to F). Single digit numbers in the hex system therefore run from 0 to F. The table gives hex to decimal to binary conversions, and should help to clarify the way in which the system operates. Note that leading zeros have not been supressed in the binary column, so that it is easier to visualise the codes on a bit by bit basis.

Hex-Decimal-Binary Conversion Chart

Hex	Decimal	Binary
0	0	0000
1	1	0001
2	2	0010
3	3	0011
4	4	0100
5	5	0101
6	6	0110
7	7	0111
8	8	1000
9	9	1001
A	10	1010
B	11	1011
C	12	1100
D	13	1101
E	14	1110
F	15	1111
10	16	10000
11	17	10001
12	18	10010

Hex switches are normally very convenient for entering header byte codes, but are more awkward for entering data bytes. However, in most cases you will not want to enter a particular data value, merely one that is very low, intermediate or high. 00 is the minimum value, 40 is an intermediate value, and 7F is the highest valid data byte value.

Output Circuit

The output direct from pin 25 of IC2 is not suitable for driving MIDI inputs. The circuit of Figure 3.4 provides the unit with two MIDI OUTs, and it consists of two identical switching circuits. TR3 operates as a standard MIDI open collector output stage, and it is preceded by an inverter stage based on TR2. This inverter is needed as TR3 would otherwise be switched on when it should be switched off, and vice versa.

Fig.3.4 The MIDI code generator output stage

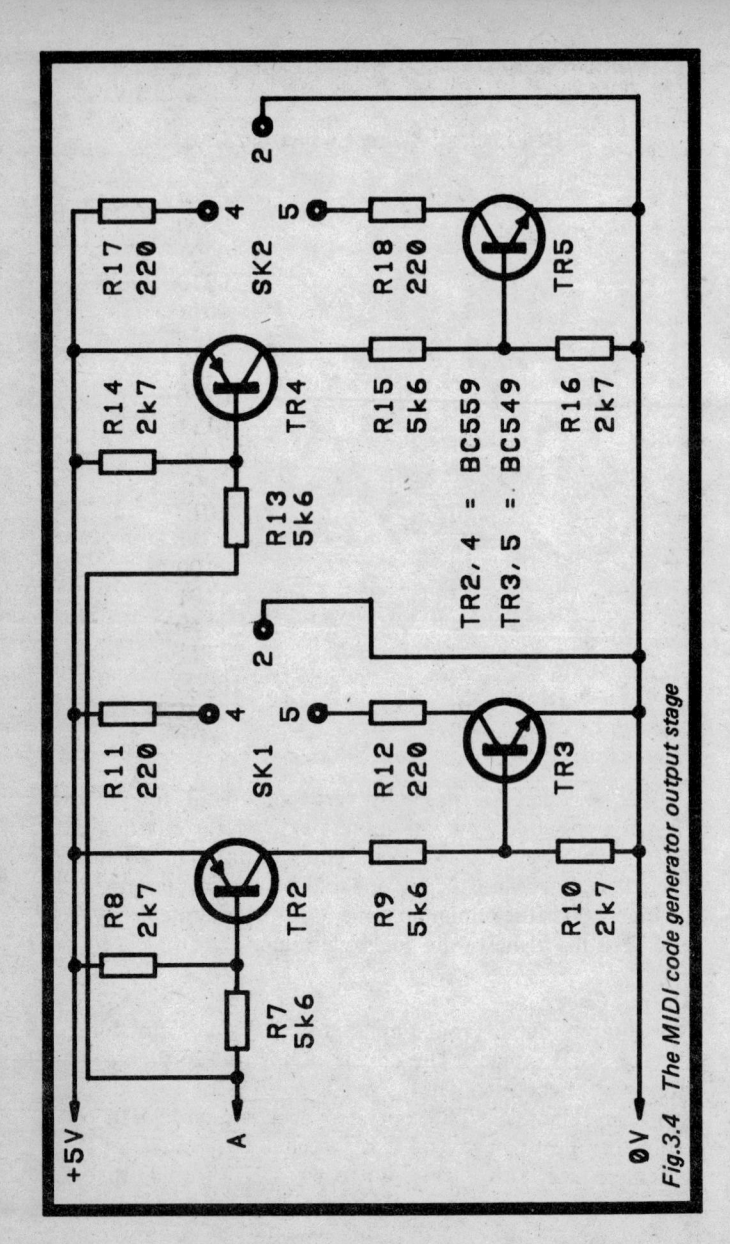

Note that if only a single MIDI output is required, R13 to R18, TR4, and TR5 should be omitted. On the other hand, if more outputs are required, further output stages could be added to the circuit. Reliable results should be obtained with four output stages, and in practice it is likely that half a dozen or more could be used without any difficulties arising. However, I have not tried the unit using more than four outputs.

Power Source

The current consumption of the circuit is approximately 15 milliamps. This is low enough to permit economic operation from batteries, and four HP7 size cells in a plastic holder are suitable. These provide a supply voltage that is a little higher than the required 5 volts, but even with fresh batteries fitted it is unlikely that the circuit would malfunction. However, four AA size nickel-cadmium rechargeable cells might be a better choice, and would give almost exactly the required supply potential of 5 volts. Another alternative is to use a 9 volt battery such as a PP9, plus a 5 volt monolithic regulator to drop the output voltage to the required level. A suitable circuit is provided in Figure 3.5.

If it is not important for the unit to be self-contained, there is the third option of a mains power supply unit. Figure 3.6 shows a suitable circuit, and this is a standard type having push-pull rectification and a 5 volt monolithic voltage regulator. FS1 should be an anti-surge type and not of the more usual quick-blow variety. The latter would tend to live up to its name and "blow" at switch-on due to the initial surge current as C1 charges up.

Of course, if you use the mains power supply it is essential to observe the usual safety precautions. The unit must be fitted in a case that has a screw fixing lid or cover, so that there is no easy way for anyone to gain access to the dangerous mains wiring. Any exposed metalwork must be earthed to the mains earth lead. What this generally means in practice is that the unit has to be fitted into a case of all metal construction. By earthing the case, any metal fixing screws etc. fitted onto it will automatically be earthed via the case. A soldertag fitted onto one of T1's mounting bolts makes a

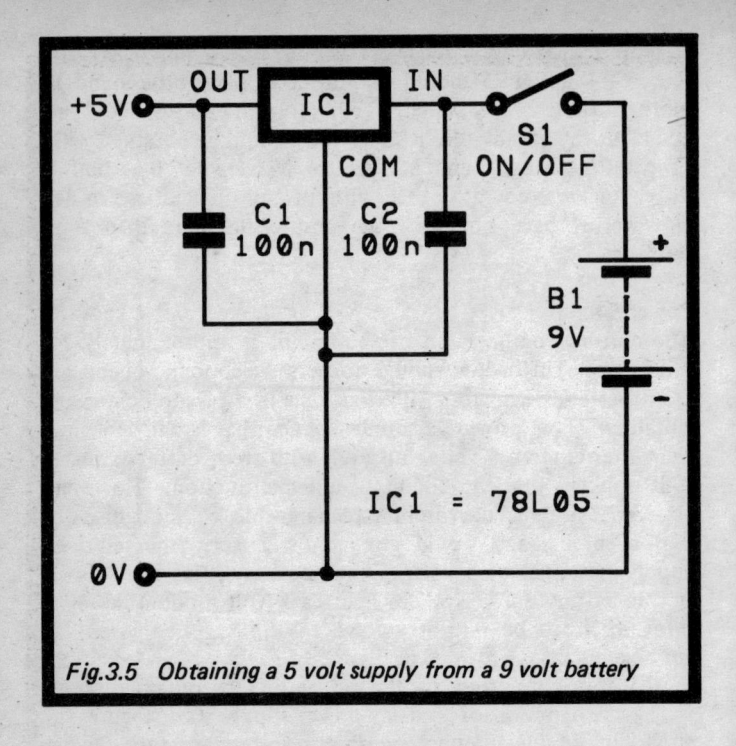

Fig.3.5 Obtaining a 5 volt supply from a 9 volt battery

convenient connection point for the mains earth lead. Needless to say, great care should be taken to avoid any mistakes in the wiring, and the finished unit should be thoroughly checked for errors before it is switched on and tested.

Construction
There should be no difficulty in building the unit using any of the standard forms of construction. Neither the 6402 or AY-3-1015D UARTs are particularly cheap, and they are both MOS devices. Consequently it would be wise to scrupulously observe the standard anti-static handling precautions when dealing with IC2. Use a 40 pin d.i.l. holder for this device, and leave it in its anti-static packaging (conductive foam, plastic tube, etc.) until it is time for it to be fitted in place. This is not until all the other components have been fitted and all the

Fig.3.6 The mains P.S.U. circuit diagram

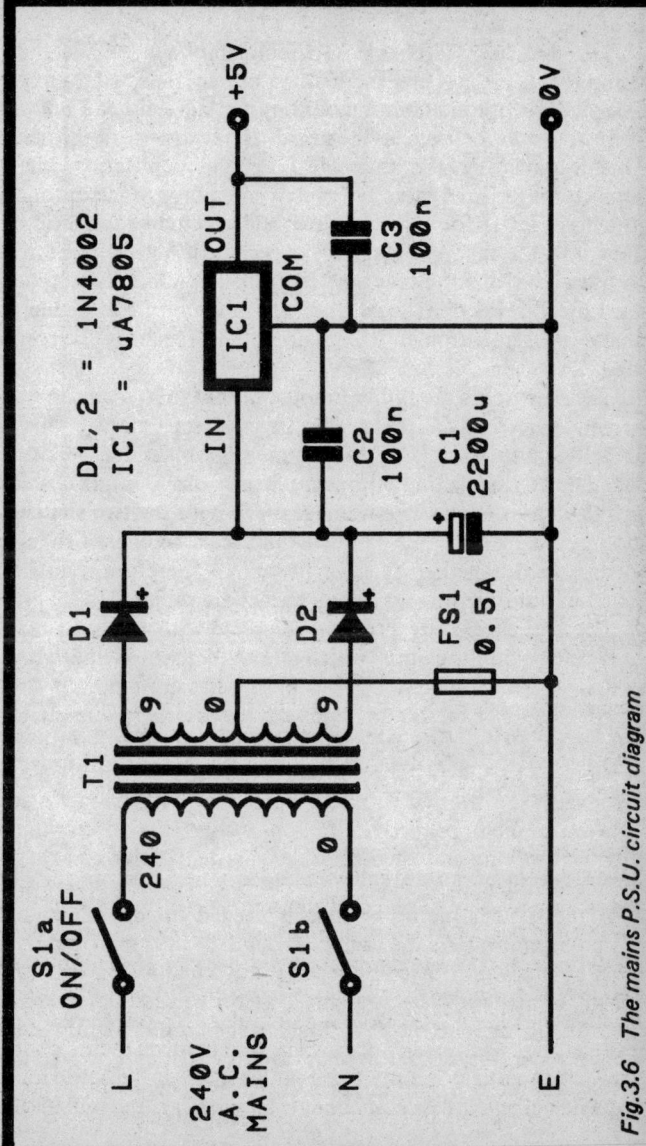

hard wiring has been completed. Handle the device as little as possible.

The two hex switches pose a minor problem. While these components are not too difficult to obtain, many of them are designed for printed circuit mounting and have either a built-in control knob or are intended for screwdriver adjustment! This is obviously less than ideal for this application where some form of panel mounted switch is required. One option is to simply settle for eight toggle or slider switches and to enter data into the unit in binary form. A much better solution is to use a pair of "thumbwheel" style hex switches. The type I used are the RS miniature type, which should be available to amateur users through RS retail outlets (such as "Electromail").

These switches are rather unusual in that they require a pair of mounting end-cheeks which are sold separately. This is actually quite sensible, as two or more switches can be fitted together as a single unit with just one pair of mounting cheeks. In this case it is clearly advantageous to have the two switches mounted close together, and it would seem to be best to have them joined together as a single unit. Therefore, two hex switches but only one set of end-cheeks are required.

The end-cheeks are supplied complete with threaded brass rods and mounting nuts which are used to hold the whole switch assembly together. The brass rods must be cut to a length that is exactly the same as the width of the switch assembly. If the rods are cut slightly too long it might not be possible to fit the switch into its mounting hole — if they are marginally too short it will not be possible to fit the mounting nuts onto them properly. It is probably best to cut them slightly too long and then carefully file them down to exactly the required length.

A rectangular mounting hole is needed for the completed switch assembly. As the switch is a click-fit into this cutout, it must be made quite accurately. It is not a major disaster if the cutout is made a little too small, as it can then be carefully filed out to the correct size. Making the cutout slightly oversize is more serious, as it may then be difficult to get the switch assembly firmly clipped in place. There would probably be no option but to use a powerful adhesive to fix the switch in position properly. The correct size for the cutout is 31 millimetres high by 24.5 millimetres. A suitable cutout can be made using a fretsaw or a coping saw, or a miniature round file can be used. Either way, it is advisable to cut the hole slightly too small and then carefully file it out to exactly the required dimensions.

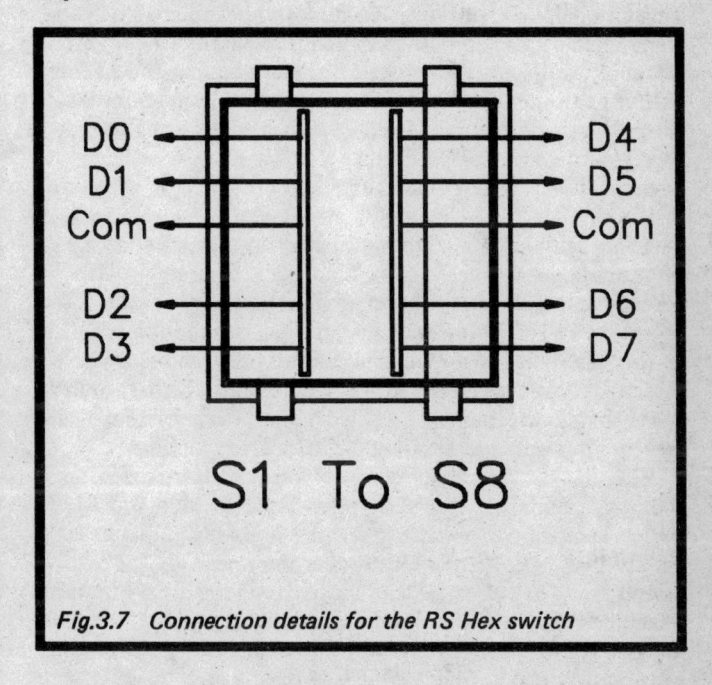

Fig.3.7 Connection details for the RS Hex switch

Connections to the switches are via a sort of miniature printed circuit board at the rear of the component. Connections can be made direct to this, but it is probably safer to make the connections via printed circuit pins. Connection details for this type of switch are provided in Figure 3.7.

Obviously the unit should operate perfectly well with other hex switches, but as pointed out previously, these are mostly intended for printed circuit mounting. Also, note that some switches are open for logic 1, and closed for logic 0. This circuit is only designed for switches that are open for logic 0, and closed for logic 1. The other type of switch can be used in this circuit, but only if the switches and load resistors are swopped over.

Components for MIDI Code Generator (Figs 3.2, 3.3 & 3.4)

Resistors (all 0.25 watt 5% carbon film)
R1	470k
R2	1k
R3	1k
R4	47
R5	1M
R6	2k2
R7	5k6
R8	2k7
R9	5k6
R10	2k7
R11	220
R12	220
R13	5k6
R14	2k7
R15	5k6
R16	2k7
R17	220
R18	220
R19 to R26	1k (8 off)

Capacitors
C1	22μ 16V elect
C2	22p ceramic
C3	22p ceramic
C4	1μ 63V elect
C5	47μ 10V elect

Semiconductors
IC1	4040BE or 74HC4040 (see text)
IC2	6402
TR1	BC549
TR2	BC559
TR3	BC549

Semiconductors (continued)
TR4   BC559
TR5   BC549

Miscellaneous
SK1   5 way (180 degree) DIN socket
SK2   5 way (180 degree) DIN socket
X1   4 MHz crystal
S1 to S4   Hex switch (see text)
S5 to S8   Hex switch
PB1   Push to make, release to break switch
16 pin d.i.l. holder
40 pin d.i.l. holder
Case, circuit board, wire, solder, etc.

Components for Mains P.S.U. (Fig.3.6)

Capacitors
C1   2200µf 16V elect
C2   100n ceramic
C3   100n ceramic

Semiconductors
IC1   µA7805 (5V 1A positive regulator)
D1   1N4002
D2   1N4002

Miscellaneous
S1   Rotary mains switch
T1   Mains primary, 9 − 0 − 9 volt 500 mA secondary
FS1   0.5 A 20 mm anti-surge
20 mm fuse holder
Circuit board, wire, etc.

Testing
The best way of test the unit is to connect its output to a computer running software which will print received MIDI

values on the screen. In the absence of a suitable system, it is just a matter of sending some trial codes/data to any MIDI device. The obvious test is to try a note on message followed by a note off type. For example, to switch on middle C on channel 0 the values sent (in hexadecimal) would be 90, 3C, 40, and to switch it off again values of 80, 3C, and 0 would be transmitted. Although some control systems that involve the transmission of multi-byte messages only work properly if there is no more than a certain gap between one byte and the next, I know of no such time limit in the MIDI system. Consequently, the inevitable delay from one byte to the next when using a unit of this type should not prevent correct operation from being obtained.

MIDI Pedal

A MIDI pedal is merely a unit that has a foot controlled switch, and which sends a program change message each time the switch is activated. The idea is to permit an instrument to be taken through several different programs (i.e. sets of sound data) without having to operate the front panel controls. There is an obvious attraction in being able to change the sound of an instrument without having to remove one hand from the keyboard. Although this is a highly useful feature, it is surprising how few instruments have any facility to use a pedal switch in this manner.

The block diagram of Figure 3.8 helps to explain the way in which this project functions. It is inevitably a bit more complicated than the MIDI code generator unit, as in this case operating the push button switch must send two bytes of information (the program change header byte and the new program number). Ideally a unit of this type would permit the user to choose a sequence of program numbers, but this feature would greatly complicate the unit. Instead, the more simple approach of having the program number start at 1, and then increment by one on successive operations of the unit is used. With most MIDI instruments it is reasonably easy to assign any desired set of sound data to each program. However, with instruments where this is not possible the pedal unit featured here will be of relatively little value.

Fig.3.8 The MIDI Pedal block diagram

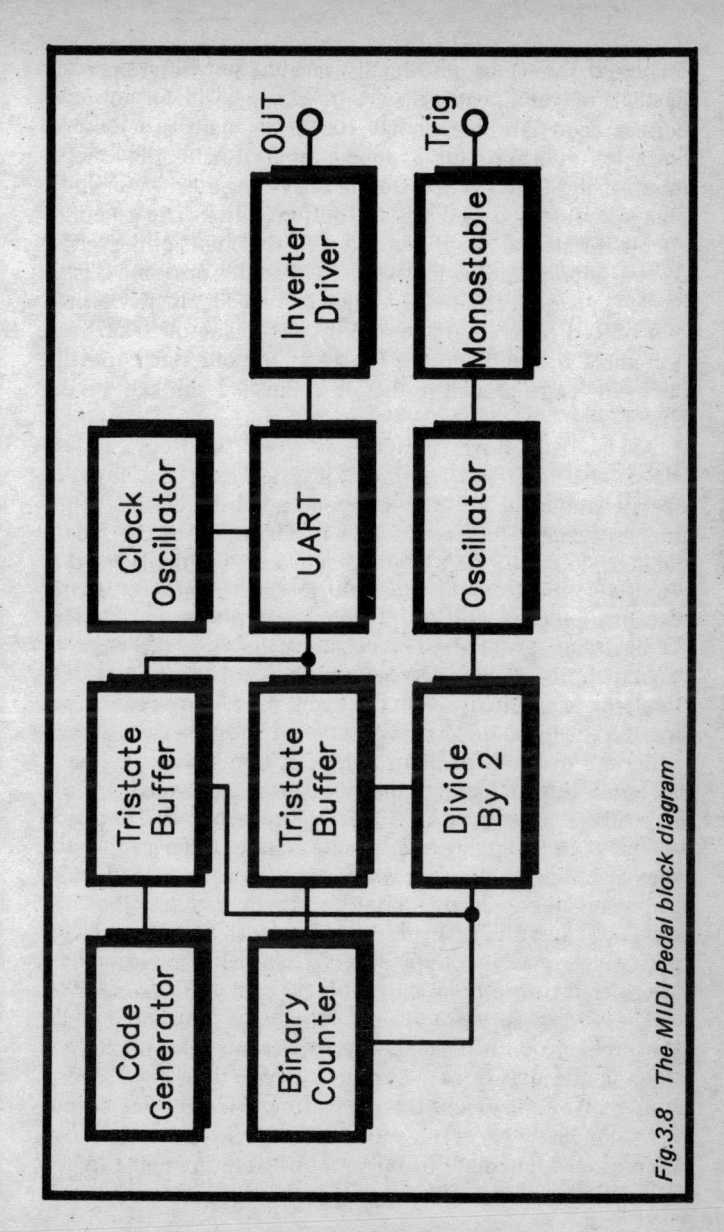

Like the previous project, this one has a UART to provide parallel to serial conversion. A clock oscillator sets the correct baud rate and an inverter/driver stage provides the correct 5 milliamp current loop output signal. The parallel input of the UART is fed from a code generator circuit and a binary counter via two tristate buffers. The code generator circuit is simply a set of wires that tie the inputs of the buffer to the appropriate logic levels, so that the program change header byte is produced. A divide by two circuit controls the two buffers, and ensures that only one or the other of them is connected through to the UART at any one time. Initially it is the code generator that is connected through to the UART.

An oscillator drives the divide by two flip/flop, and also triggers the UART. Therefore, if the oscillator is permitted to operate normally, the unit will send a continuous stream of program change messages. The oscillator drives the binary counter via the divide by two circuit, and it is the latter that sets the program number on each transmitted message. Therefore, the program number increments in the desired fashion. As described so far the circuit almost provides the required function, but obviously some means of controlling the oscillator is required, so that the unit sends program changes one at a time, and under manual control.

Several methods of controlling the circuit were tried, and the most simple method that worked reliably was to use a monostable multivibrator. This controls the oscillator, and normally holds it in the inactive state. Triggering the monostable using the foot switch results in an output pulse that gates the oscillator briefly into operation. Provided the pulse duration is made roughly equal to two oscillator clock cycles, a two byte program change message will be transmitted before the oscillator is cut off by the monostable.

There is an apparent flaw in this system in that it could give problems with "creeping sync". If the pulse duration is other than precisely two oscillator cycles, after a few operations of the foot switch the unit might produce three output bytes, or just one. The more accurate the pulse duration, the greater the number of operations before the unit glitches. In practice this does not happen, as the oscillator commences

Fig.3.9 The main MIDI Pedal circuit diagram

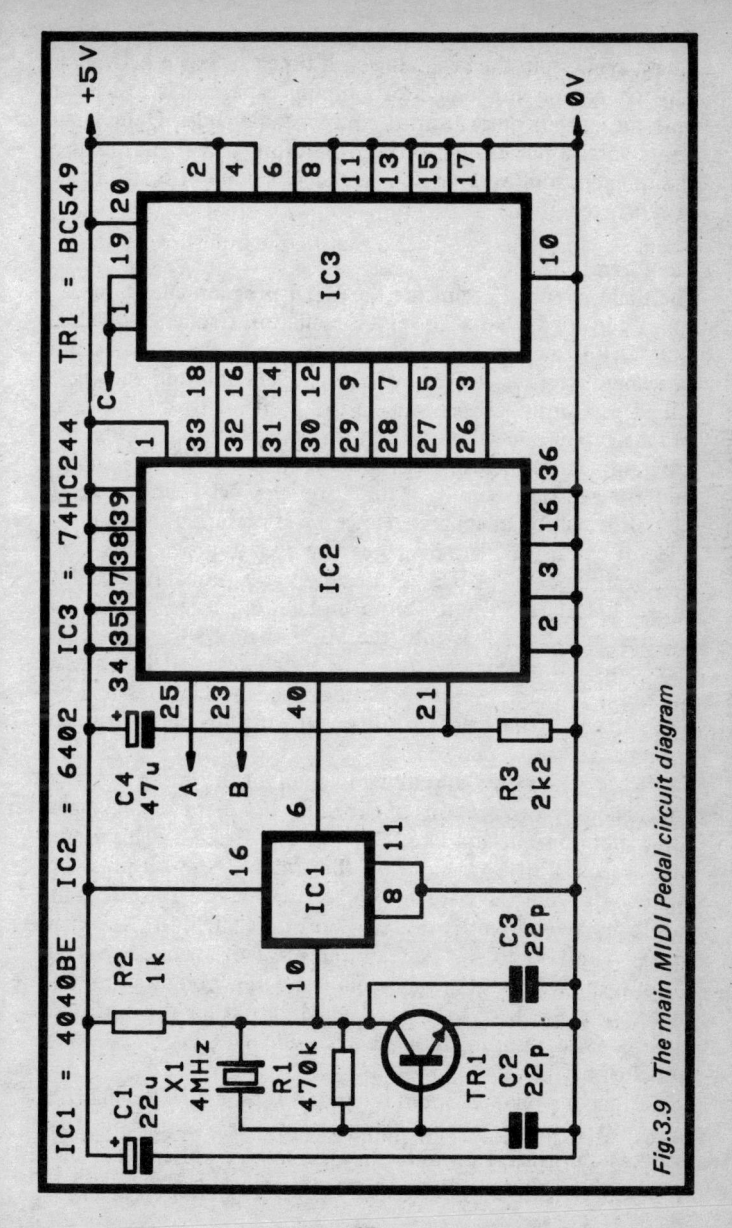

a new cycle from the beginning each time the unit is activated. Any charge on the oscillator's timing capacitor is lost each time the output pulse from the monostable ends. Quite large errors in the pulse duration are therefore acceptable, as the circuit incorporates what is a crude but quite effective form of synchronisation.

The Circuit

The main circuit diagram for the MIDI program change pedal appears in Figure 3.9. The clock oscillator, frequency divider, and UART sections of the unit are much the same as the equivalent sections of the MIDI code generator project described previously. IC3 is one of the eight bit tristate buffers, and its inputs are wired to give the correct header byte for program change messages (11000000 in binary). This assumes that transmission is only required on channel 1, and in most cases this will be adequate. However, if preferred the channel selector circuit of Figure 3.10 can be used to provide the least significant nibble for IC3. A hex switch is probably the best choice for S1 to S4, but a little thought is needed in order to convert switch settings into the MIDI channels they provide. The switch is calibrated from 0 to F, whereas MIDI channels run from 1 to 16. Some recalibration of the switch would be more than a little helpful, but might prove to be difficult in practice.

Figure 3.11 shows the circuit diagram for the binary counter and divide by two sections of the unit. IC4 is the second eight bit tristate buffer, and like IC3, it is a 74HC244. This is the high speed CMOS version of the 74244. Its outputs are switched to the on state by taking pins 1 and 19 low. The divide by two circuit is one section of a CMOS 4013BE dual D type flip/flop (IC6). Its Q output drives IC3 while its not Q output drives IC4. This may seem to be the wrong way round, but before the first byte has been transmitted the flip/flop changes state, resulting in IC3 being switched on and IC4 being switched off.

IC5 is the binary counter, and this is a CMOS 4024BE seven stage binary type. Seven stages are adequate in this application, bearing in mind that the most significant bit of a data byte is always set to zero. The clock input of IC5 is

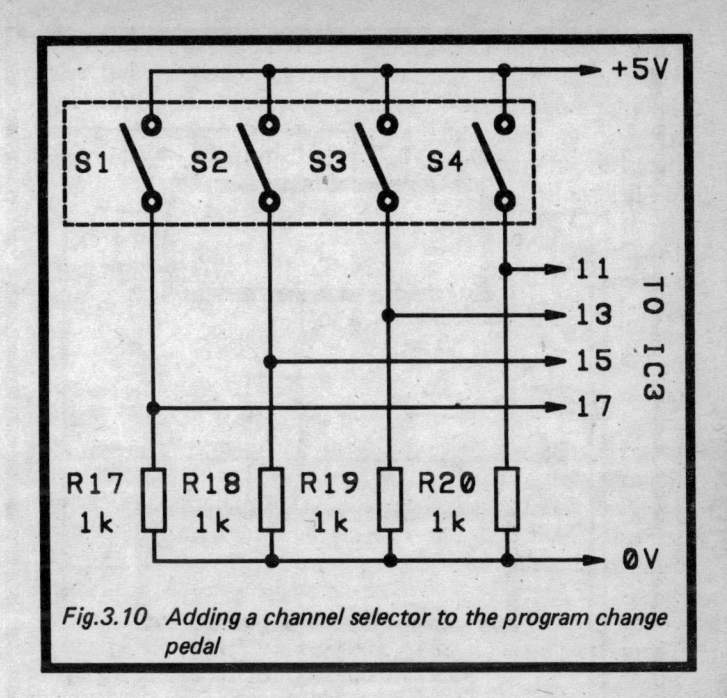

Fig.3.10 Adding a channel selector to the program change pedal

driven from the not Q output of IC6. This results in IC5 being incremented from 0 to 1 by the time the first data byte is transmitted. The first operation of the unit therefore sets the controlled unit or units to program 1. This is not really a fault, since the controlled equipment will presumably be set to program 0 initially, and the unit is then required to move it to program number 1 on the first operation of the foot switch. Note though, that you can start from any desired program number, but that the pedal unit will always take the controlled equipment to program 1 on the first operation, program 2 on the next operation, and so on. C5 and R4 provide an initial reset pulse to IC5 and IC6 at switch-on. PB1 can be used to manually reset the unit whenever necessary.

The circuit diagram for the monostable and oscillator stages of the unit appear in Figure 3.12. Both stages are based on standard 555 circuits, but in order to keep the current

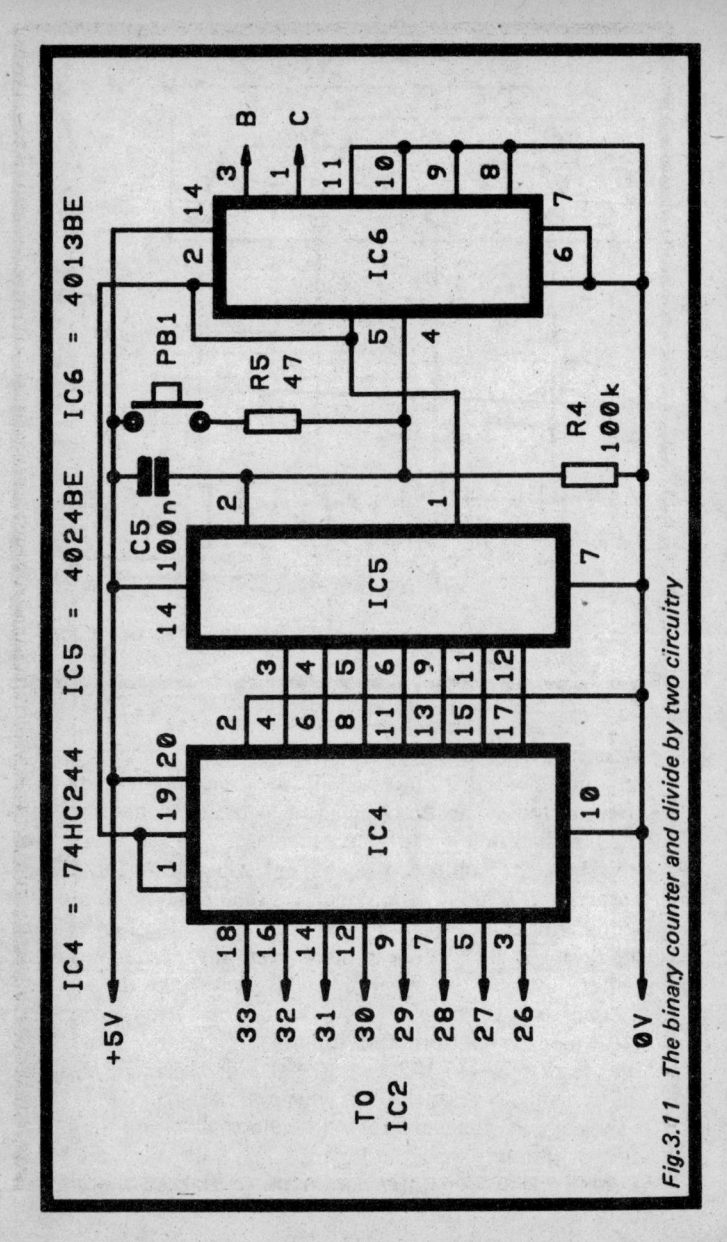

Fig.3.11 The binary counter and divide by two circuitry

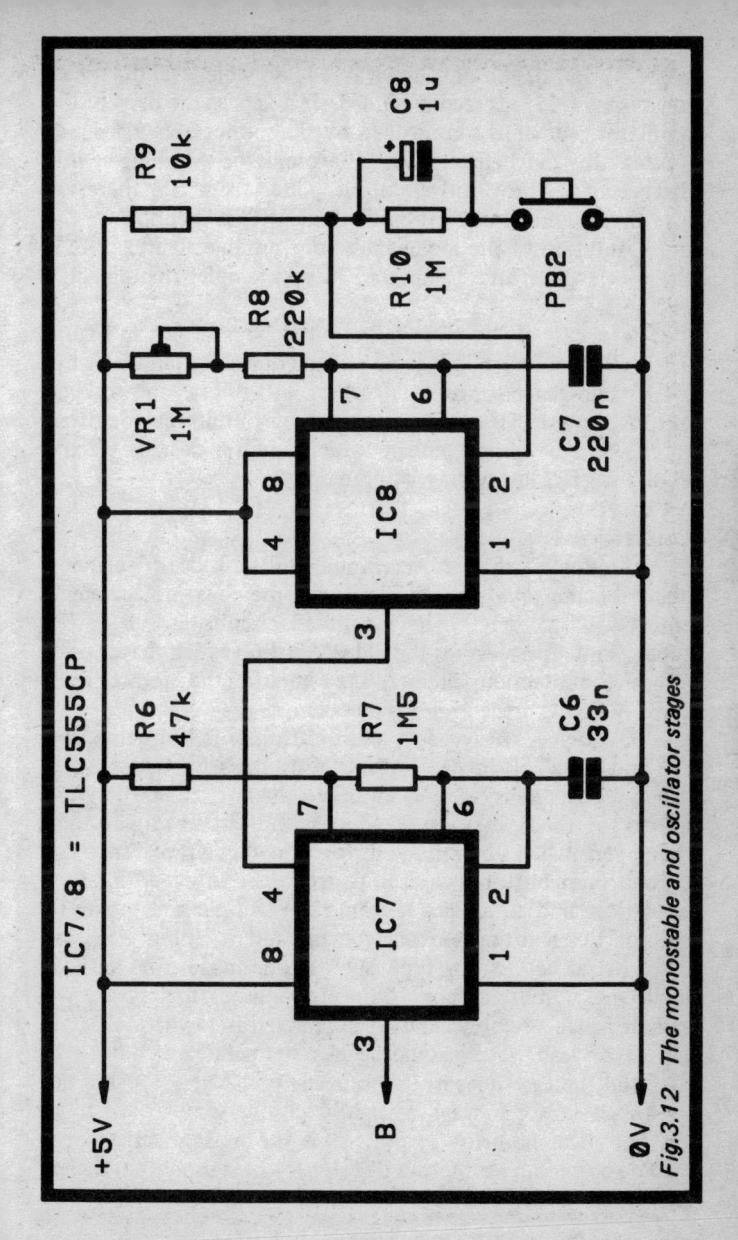

Fig.3.12 The monostable and oscillator stages

consumption of the unit down to a reasonable level the use of low current 555s is recommended. IC7 acts as the basis of the oscillator, and this is controlled by the monostable via pin 4. Taking this pin high activates the oscillator — taking it low disables it. The positive output pulses from IC8 therefore give the required bursts of oscillation. VR1 is used to trim the pulse duration of the monostable to a suitable figure. PB2 is the foot switch, and it triggers IC8 via a simple "debouncing" circuit.

The circuit of the inverter/output stage is shown in Figure 3.13. This is much the same as the equivalent circuit in the MIDI code generator unit. However, in this case only a single output is used. Of course, two or more outputs can be fitted to the unit if desired, and it is just a matter of adding some extra inverter/output stages to the unit.

Construction

Construction of this project should not be too difficult, and many of the points raised about the previous project apply equally to this one. The integrated circuits are all CMOS types, and apart from the TLC555CPs (which have fully effective protection circuits) they require the normal anti-static handling precautions to be taken.

PB2 can be a heavy duty push button switch mounted on the top panel of the case. Heavy duty push button switches can be rather noisy and awkward to operate. Also, they are mostly of the latching type, whereas this circuit requires one that opens when you raise your foot. You might prefer to use a large push button switch not specifically intended for foot operation, and to replace it from time to time if it proves to be not quite strong enough for the task. Either way, the case should be a strong type, such as one made from steel or a diecast aluminium box. An alternative is to buy a foot-switch/pedal assembly, and to connect this to the main unit via a twin lead. This should be easy to operate and reliable in use, and the case does not then need to be a very tough type. Although in some ways a slightly cumbersome solution to the problem, in many respects it is the most practical one.

The current consumption of this project is somewhat higher than that of the MIDI code generator unit, but the same

Fig.3.13 The MIDI pedal output stage

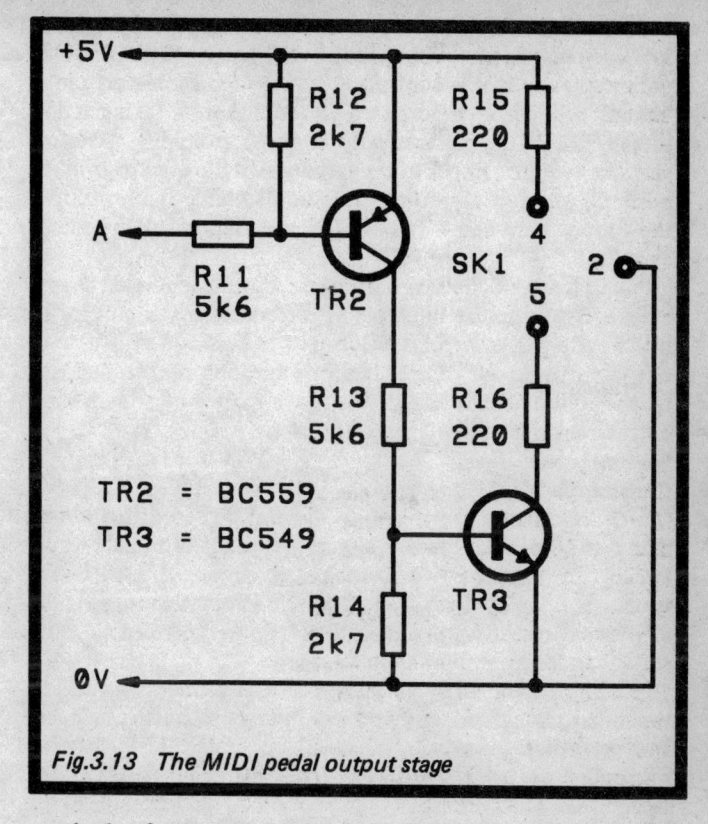

methods of powering it apply. With a unit of this type portable operation is often a desirable asset, and batteries may well be the most convenient method of powering the unit.

Components for MIDI Pedal (Figs 3.9, 3.11, 3.12 & 3.13)

Resistors (all 0.25 watt 5% carbon film)
R1	470k
R2	1k
R3	2k2
R4	100k
R5	47

Resistors (continued)

R6	47k
R7	1M5
R8	220k
R9	10k
R10	1M
R11	5k6
R12	2k7
R13	5k6
R14	2k7
R15	220
R16	220

Potentiometer

VR1	1M sub-min hor preset

Capacitors

C1	22μ 16V elect
C2	22p ceramic
C3	22p ceramic
C4	47μ 10V elect
C5	100n polyester
C6	33n polyester
C7	220n polyester
C8	1μ 63V elect

Semiconductors

IC1	4040BE or 74HC4040
IC2	6402
IC3	74HC244
IC4	74HC244
IC5	4024BE
IC6	4013BE
IC7	TLC555CP or similar
IC8	TLC555CP or similar
TR1	BC549
TR2	BC559
TR3	BC549

Miscellaneous

SK1	5 way (180 degree) DIN socket

Miscellaneous (continued)
PB1   Push to make, release to break switch
PB2   Heavy duty push to make, release to
                break switch
X1   4 MHz crystal
8 pin d.i.l. holder (2 off)
14 pin d.i.l. holder (2 off)
16 pin d.i.l. holder
20 pin d.i.l. holder (2 off)
40 pin d.i.l. holder
Case, circuit board, wire, etc.

Additional Components for Channel Selector (Fig.3.10)

Resistors (all 0.25 watt 5% carbon film)
R17 to R20   1k (4 off)

Miscellaneous
S1 to S4   Hex switch

Setting Up
Setting up the unit is easiest if a computer can be used to monitor the output from the unit. It is then just a matter of using trial and error to find a setting for VR1 that results in each operation of the foot switch instigating the transmission of two bytes. In the absence of a computer and software to monitor the output of the unit, it can be connected to any MIDI device that responds to program change messages, and trial and error can again be used to find a setting for VR1 that gives satisfactory results. There should be a range of settings that give correct operation of the unit, and VR1 should be set at roughly the middle of this range.

MIDI Programmer
Although modern electronic musical instruments are undoubtedly much more sophisticated than those of a decade ago, there is one respect in which most users seem to consider them substantially inferior. This is the ease

(or difficulty) with which the sound generator circuits can be adjusted. In the days of analogue synthesisers there was a separate and clearly marked control knob for each parameter that could be adjusted. Altering and setting up sounds was very easy because you could adjust any parameter immediately simply by grabbing the correct control knob.

This approach does not lend itself too well to modern digital instruments where there are generally a lot more parameters to adjust, with a different set of adjustments for each of what is usually six to sixteen voices. A separate control for each parameter would probably cost hundreds of pounds to implement, possibly even thousands! This would result in instruments that were excessively expensive and were physically huge. In the early days of digitally controlled instruments a popular solution to the problem was to have a single control knob, with a keypad being used to assign this control to whatever parameter you wished to adjust. A numeric display usually indicated the number of the current control, and in conjunction with a chart it was reasonably easy to ascertain that you were adjusting the correct parameter.

In my experience this system worked quite well. It was not as good as having a separate control for each function that needed to be adjusted, but under the circumstances it gave a good compromise between size and cost on the one hand, and ease of use on the other.

Unfortunately, this system seems to have given way to one which dispenses with the control knob altogether. These days the standard system is to have a keypad, and as before, this is used to select the desired parameter. Also as before, a display normally shows you which parameter has been selected. Things are actually a bit more advanced in this respect, in that a modern display is often an alpha-numeric type which will tell you which parameter you are adjusting (e.g. "Filter Envelope Attack") rather than just showing the number of the control. The parameter is usually adjusted by entering new values via the keypad, or perhaps by using "up" and "down" keys to increment or decrement the control's setting.

This system may be popular with instrument manufacturers, but it is not one that is favoured by many users. Setting up and "fine tuning" the sound generator circuits tends to be a very long winded process indeed. This fact, plus the complexity of modern instruments, has led to many users giving up any ideas of setting up their own sounds, and simply settling for the factory presets or other ready programmed sounds.

With some instruments MIDI offers the opportunity to go back to the convenience of the control knob era. Where an instrument allows the sound generator circuits to be adjusted via MIDI controller messages, an external unit with a set of control knobs can be used to program the sound generator circuits. Such a unit is likely to be quite expensive though. This leads us back to the single knob compromise, where one control is used for every parameter, with some means of selecting the control you wish to adjust. This is something that can be easily implemented using quite a simple MIDI controller unit.

The unit featured here uses two hex switches to select the required MIDI control number, and has a control knob which can be used to vary that controller over its full 0 to 127 data range. It gives easy control of any MIDI controller, including switch types which are easily set on or off by adjusting the control fully in one direction or the other.

It is only fair to point out that it is a good idea to check the MIDI specifications of your equipment before building a unit of this type. The specification might show that there is easy access to a wide range of parameters via MIDI controllers. There might even be a facility that enables each controllable feature to be assigned to a user specified MIDI control number. On the other hand, some recent instruments permit only a few basic functions such as modulation depth and master volume to be accessed via MIDI controllers. There may be no MIDI access to such things as the filters and envelope shapers, or (more probably) they might only be accessible via system exclusive messages. Obviously a MIDI controller of the type described here is of relatively little use with a unit that falls into this second category.

System Operation

The requirements for a MIDI controller are somewhat different to those for the MIDI pedal described previously. Whereas

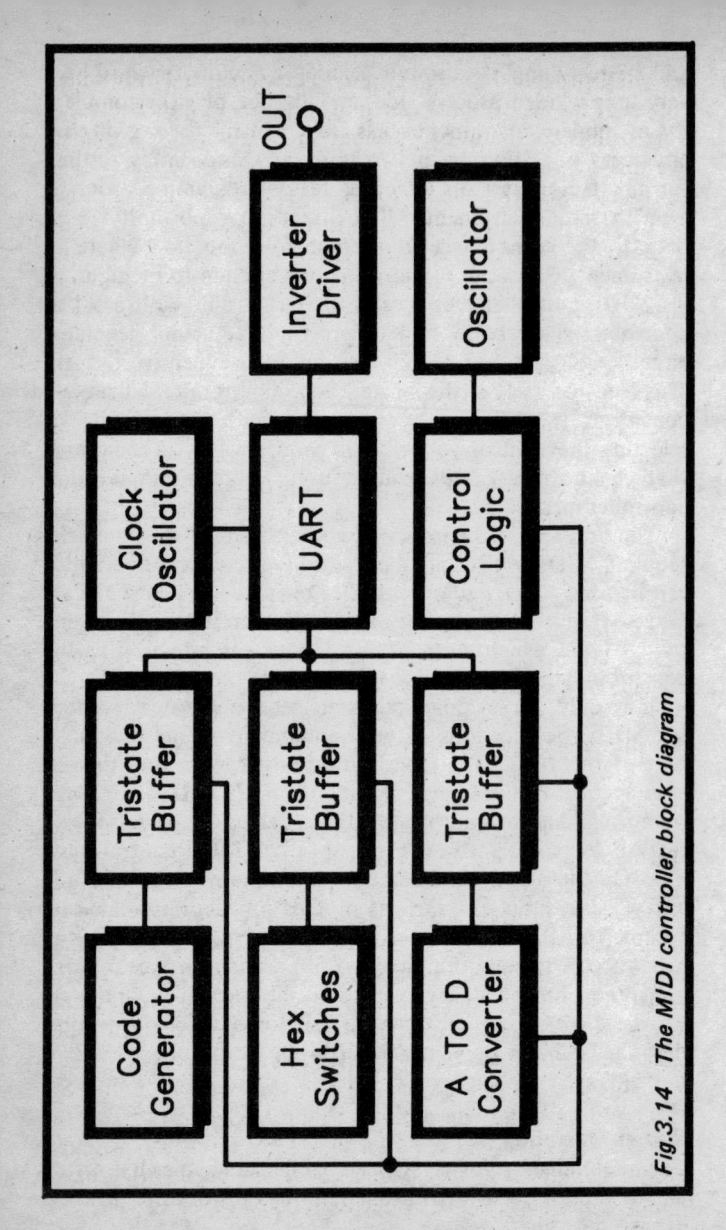

Fig.3.14 The MIDI controller block diagram

the MIDI pedal had to send two byte messages, and only when triggered, the MIDI controller must send three byte messages more or less continuously. The modified arrangement used to give this action is shown in the block diagram of Figure 3.14.

The clock, UART, and output stages are the same as those of the MIDI pedal unit. This unit also has the parallel input of the UART fed via tristate buffers, but in this case there are three of them, as each message contains three bytes. The first tristate buffer is fed from a code generator circuit which simply ties the inputs of the buffer to the appropriate logic levels so that the correct header byte is produced. The second tristate buffer is fed from a hex switch, and this is used to select the required MIDI control number. An analogue to digital converter feeds the input of the third tristate buffer, and this circuit is fed with the output voltage of a potentiometer. This component is, of course, the control that is used to vary the MIDI controller value.

A control logic circuit ensures that each of the tristate buffers are activated in turn, and that the UART is triggered into action as each new byte of data is fed to its parallel inputs. An oscillator controls the rate at which data is transmitted, and a frequency of several hundred hertz ensures that there is no appreciable delay between adjusting a control and the MIDI equipment responding to the change in value.

The Circuit

The circuit diagram for the clock and UART sections of the controller appears in Figure 3.15. This requires little comment as it is much the same as the equivalent section of the program change pedal project. The only difference is that the wiring at the input to IC3 has been changed to give the control change header byte (10110000 in binary). The circuit only transmits on MIDI channel 1, but like the program change pedal, a hex switch circuit can be used to provide the least significant nibble for IC3 if multi-channel operation is required.

Figure 3.16 shows the control number selector circuit. S1 to S3 and S4 to S7 will presumably be a pair of hex switches. In the case of S1 to S3 the most significant bit of the switch is left unconnected, so that values above the legal MIDI maximum (7F hex) can not be accidentally selected.

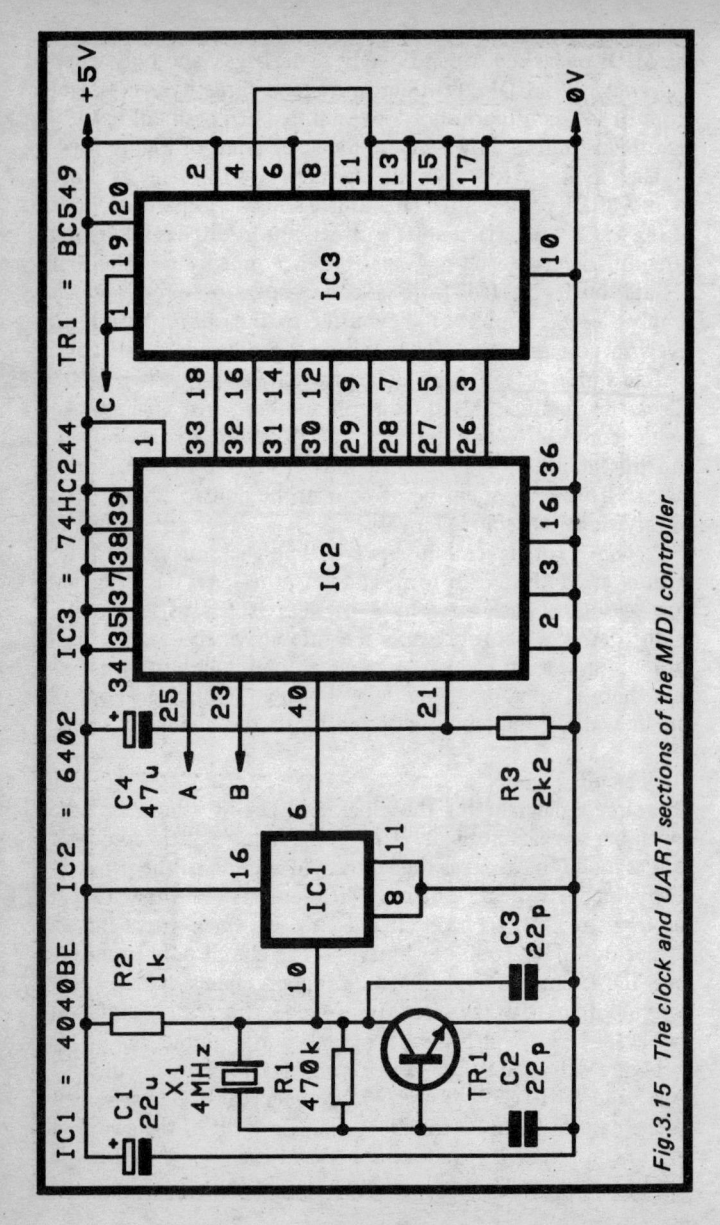

Fig.3.15 The clock and UART sections of the MIDI controller

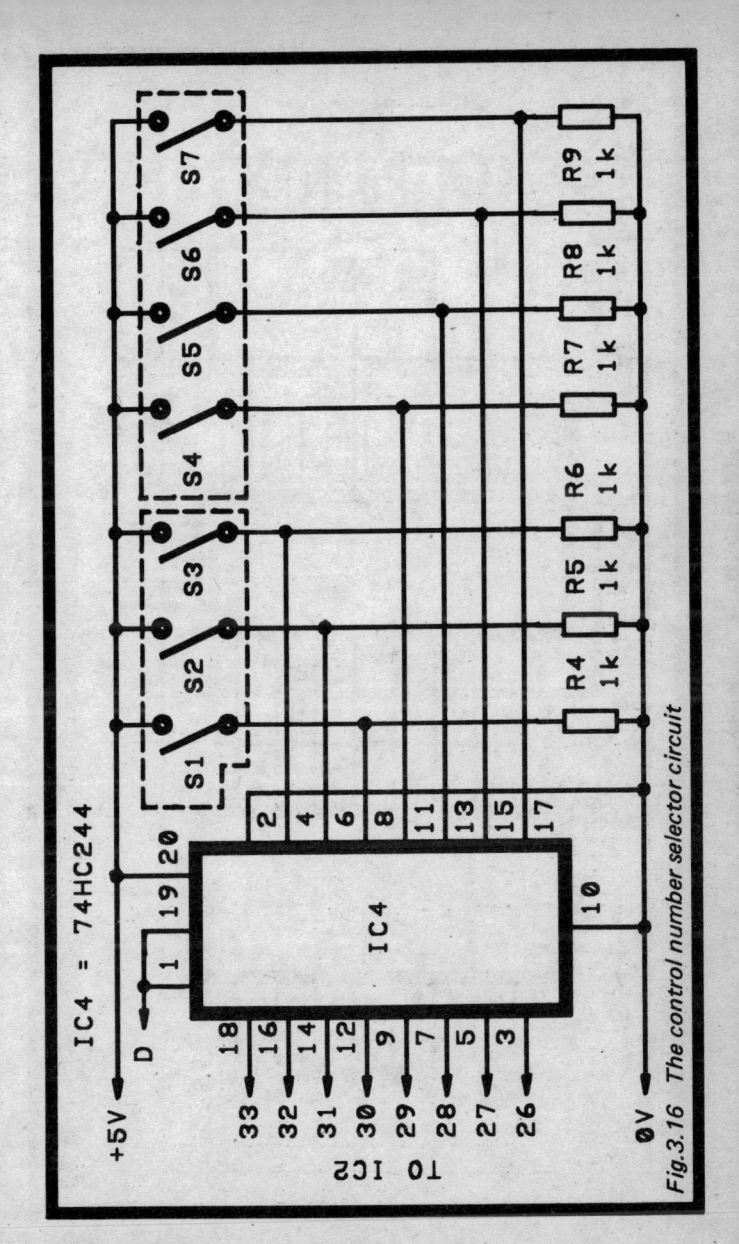

Fig.3.16 *The control number selector circuit*

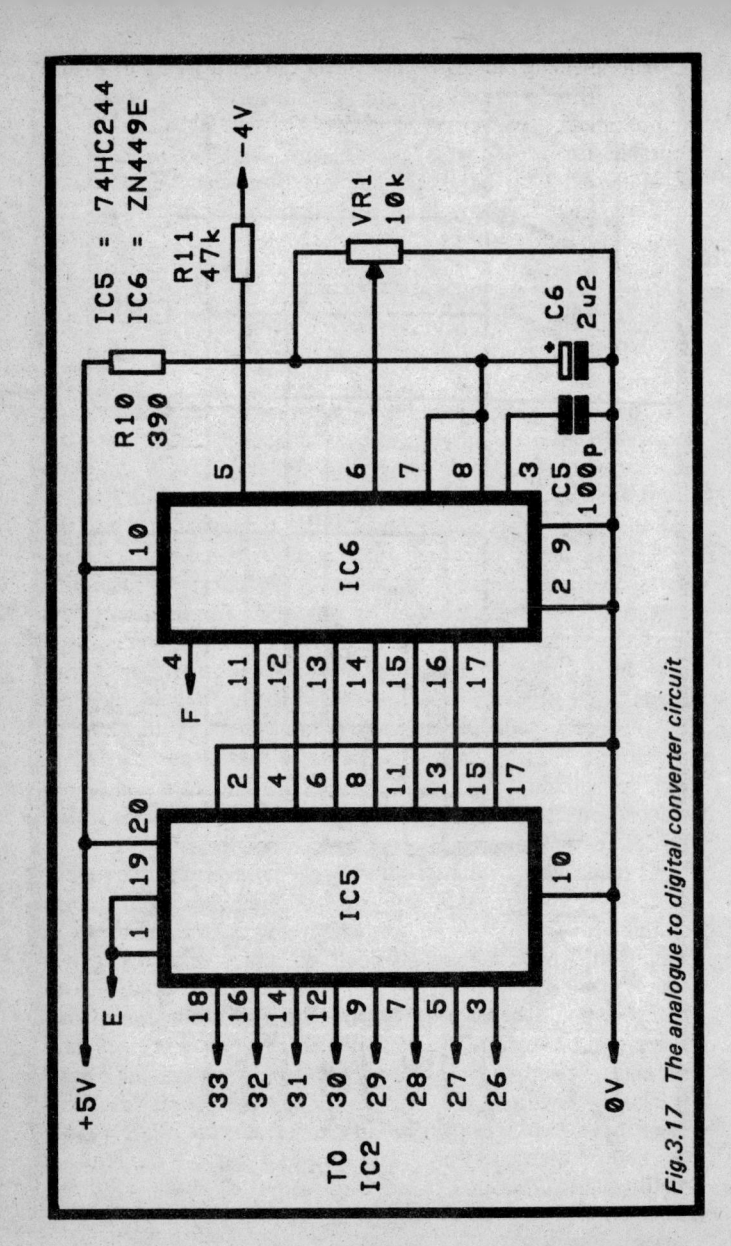

Fig.3.17 The analogue to digital converter circuit

The analogue to digital converter circuit appears in Figure 3.17. This is based on the Ferranti ZN449E successive approximation converter. The ZN447E and ZN448E are also suitable for use in this circuit, and they differ from the ZN449E only in that their guaranteed accuracy is somewhat better. High accuracy is of little importance in this application, and the lower cost of the ZN449E makes it a better choice. However, it is a bit more difficult to obtain than the ZN447E and ZN448E, and it might be necessary to use one of these if the ZN449E proves to be elusive.

VR1 is the control potentiometer, and it is fed from IC6's internal 2.55 volt regulator. R10 and C6 are the load resistor and decoupling capacitor for this regulator. C5 is the timing capacitor in IC6's internal clock oscillator, and it sets the clock frequency at typically a little under 1 MHz. It takes no more than nine clock cycles per conversion, or about 10 μs in other words. This is obviously more than adequate for the present application, where only a few hundred conversions per second are required. IC6 is an 8 bit converter, but in this case only a seven bit output is required. Consequently, the least significant output of IC6 is left unused. IC6 requires a negative "start conversion" pulse in order to trigger it into action, and this is provided by the control logic stage while the header byte is being transmitted. There is an "end of conversion" status output available, but it is not needed in this application. Each conversion will have been completed well before the output of the converter is fed through to the UART for transmission.

The analogue to digital converter requires a negative supply of about 4 volts for the "tail" resistor in its high speed voltage comparator (R11). If the unit is powered from a mains power supply it would not be difficult to add a simple negative supply generator circuit to it. With battery power extra batteries could be used to provide the negative supply. However, in either case it is probably easier and better to generate the negative supply from the +5 volt supply. A suitable circuit appears in Figure 3.18. This is based on a couple of CMOS inverters which are utilized in the standard CMOS astable (oscillator) configuration. The output from the oscillator is rectified and smoothed to give an output of about −4 volts.

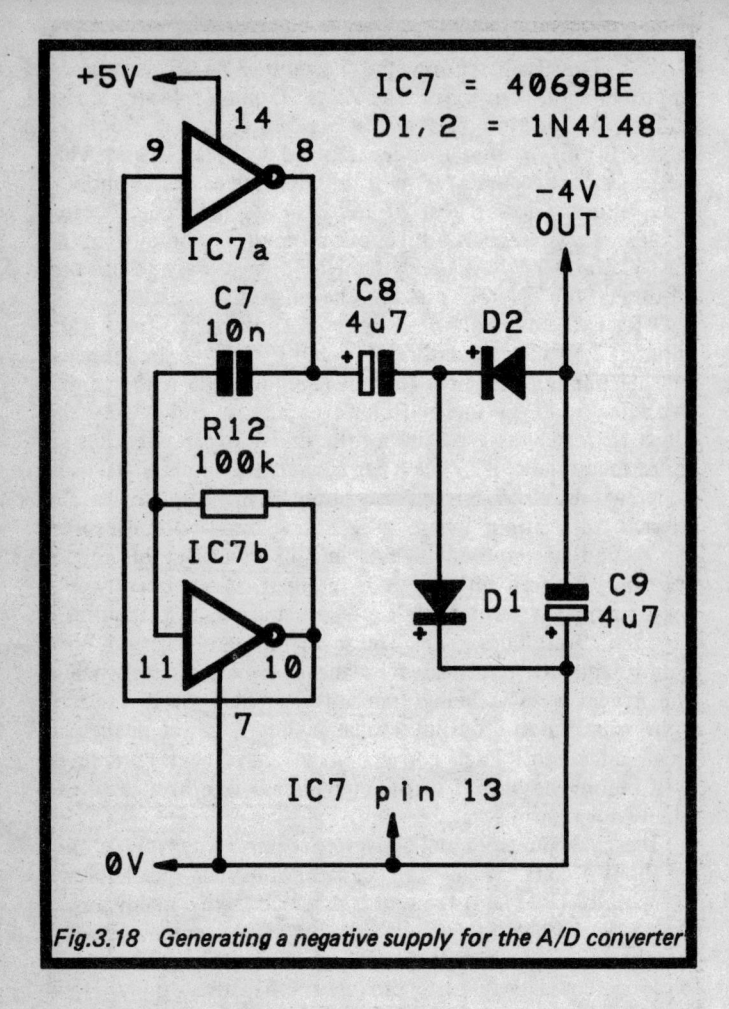

Fig.3.18 Generating a negative supply for the A/D converter

Only a very limited supply current is available from the output of this circuit, but as the ZN449E only draws about 60 microamps from its negative supply there is no problem in this respect.

Note that the two inverters are "spares" from the control logic circuit, which uses a further three of the six inverters in

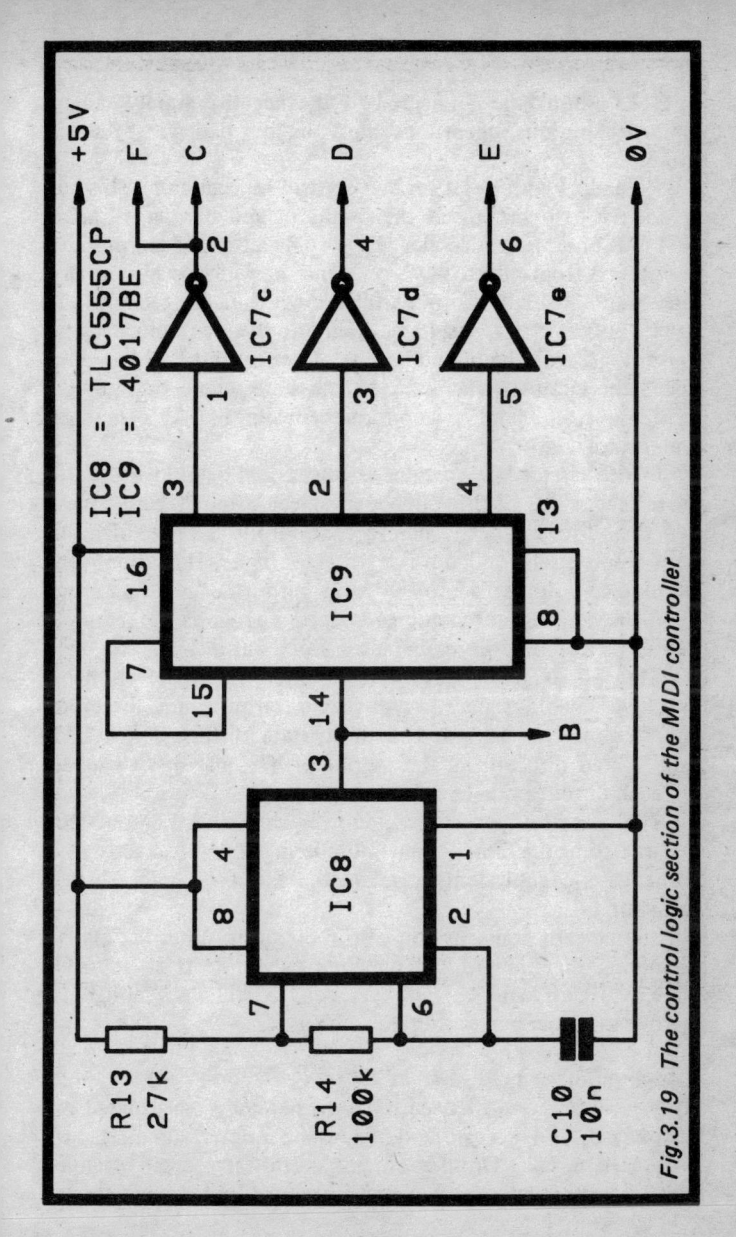

Fig.3.19 The control logic section of the MIDI controller

IC7. This leaves one inverter unused, and its input terminal (pin 13) should be connected to the negative supply rail to prevent spurious operations and possible damage by static charges.

Refer to Figure 3.19 for the control logic circuit. This has a 555 (IC8) operating as the oscillator and driving a CMOS 4017BE one of ten decoder (IC9). The latter has ten outputs numbered from "0" to "9", only one of which is high at any one time. Initially output "0" is high, but after one input cycle output "1" goes high, then on the next clock cycle output "2" goes high, and so on. After output "9" has gone high, the circuit cycles back to the state where output "0" is high again, and it continues cycling in this manner indefinitely.

In this circuit IC9 operates as what could be more accurately termed a one of three decoder, since output "3" is connected to the reset input. Therefore, as output "3" goes high the unit resets itself, and output "0" goes high. This effectively eliminates outputs "4" to "9", and output "3" does no more than provide this automatic resetting. This gives the required sequence of three control pulses from outputs "0" to "2", but the pulses are positive whereas the tristate buffers need negative enable signals. All three outputs are therefore inverted prior to being fed to the tristate buffers. The UART is triggered direct from the output of IC8, and it is triggered on each output cycle from IC8.

The invert/output stage circuit is shown in Figure 3.20. Apart from the component numbering this is exactly the same as the equivalent stage of the MIDI program change pedal unit.

The current consumption of this circuit is somewhat higher than that of either of the previous two projects at approximately 50 milliamps. However, this is still low enough to permit the same power sources to be used.

Construction and Use
This project is a bit too complex to be easily constructed on stripboard, and a custom double-sided printed circuit board would be ideal. On the other hand, do-it-yourself double-sided boards are far from easy to produce, and either a single-

Fig.3.20 The inverter/output stage

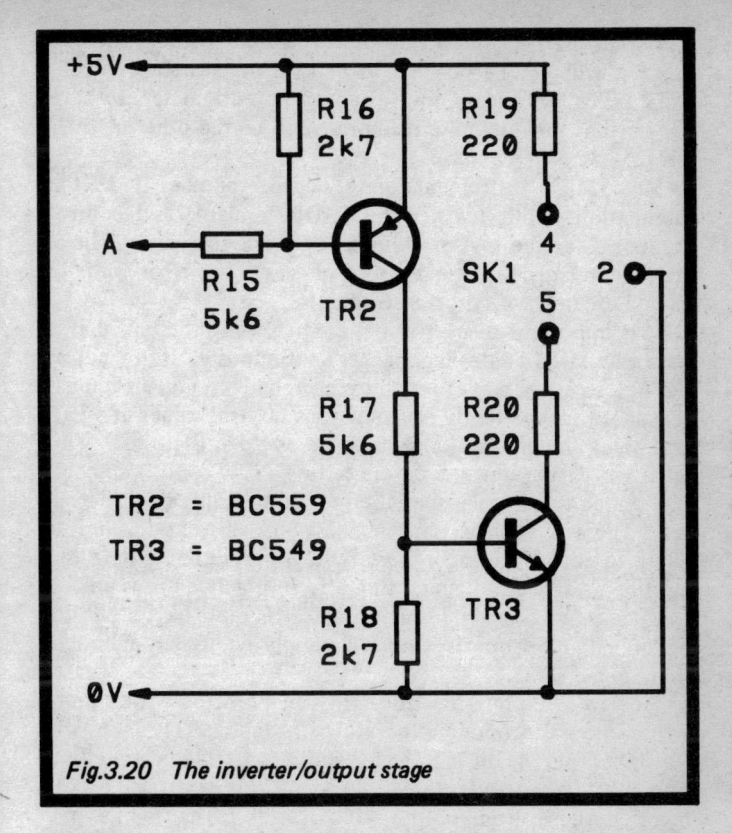

sided board or a stripboard plus a substantial number of link wires are probably more practical propositions. Stripboard probably represents the most simple method of construction, but is only suitable if the unit does not need to be particularly small. As MIDI units are often housed in 19 inch rack-mount cases, small size will probably not be a major consideration in most instances.

Use a good quality component for VR1. Apart from the fact that this control is likely to receive a lot of use and may quickly wear out if it is of mediocre quality, cheap types may fail to operate properly at all. It is important that the wiper of VR1 can reach both ends of its track, so that the full 0 to

127 control range is available from the unit. Without this full range it will not be possible to control the switch type controls. It is not uncommon for cheap potentiometers to have wipers that do not quite reach one end or the other of their tracks.

Most MIDI instruments have control number 1 as the modulation depth (as per the MIDI standard) and control number 7 as the master volume (as per a sort of unofficial MIDI standard). These provide an easy means of initially testing the unit with most instruments.

It is important to bear in mind when using the unit that it transmits MIDI controller changes continuously. If you adjust the hex switches while the unit is switched on, you are almost certain to inadvertently set numerous control values at whatever value the control happens to be set to at that time! You must switch the unit off before setting a new control number, and switch it on again once the new control number has been set. There is a slight flaw in doing this, in that the unit is likely to cut off during a three byte group, rather than finishing the message and then cutting off. In practice this is unlikely to cause any major problems, and it is quite likely that the equipment fed from the unit will simply ignore any messages that are cut short.

However, if preferred, the unit can be fitted with a mute switch using the modification shown in Figure 3.21. Normally Ra takes pin 4 of IC8 high, and the oscillator functions normally. When Sa is closed, pin 4 of IC8 is controlled by the inverted signal from output "0" of IC9. This is high and enables the circuit to function normally until the start of a new message when output "0" goes high. The circuit is then brought to a halt before the new header byte is transmitted, and whole three byte messages are always transmitted.

Components for MIDI Programmer
(Figs 3.15, 3.16, 3.17, 3.18, 3.19 & 3.20)

Resistors (all 0.25 watt 5% carbon film)
R1   470k
R2   1k
R3   2k2

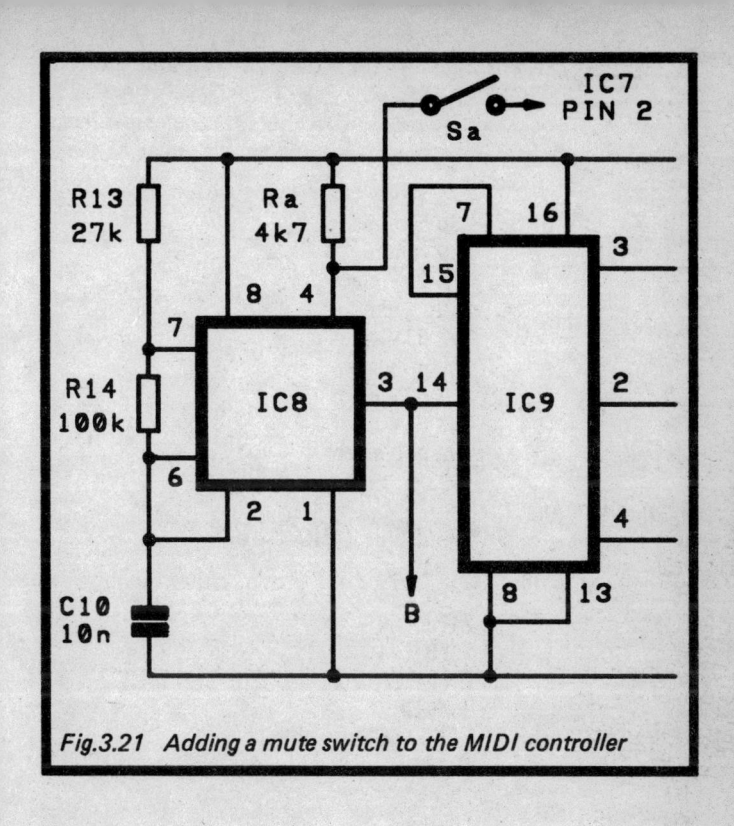

Fig.3.21 Adding a mute switch to the MIDI controller

Resistors (continued)

R4 to R9	1k (7 off)
R10	390
R11	47k
R12	100k
R13	27k
R14	100k
R15	5k6
R16	2k7
R17	5k6
R18	2k7
R19	220
R20	220

Potentiometer
VR1   10k lin

Capacitors
C1   22µ 16V elect
C2   22p ceramic
C3   22p ceramic
C4   47µ 10V elect
C5   100p ceramic
C6   2µ2 63V elect
C7   10n polyester
C8   4µ7 63V elect
C9   4µ7 63V elect
C10   10n polyester

Semiconductors
IC1   4040BE or 74HC4040
IC2   6402
IC3   74HC244
IC4   74HC244
IC5   74HC244
IC6   ZN449E (or ZN447E or ZN448E)
IC7   4069BE
IC8   TLC555CP or similar
IC9   4017BE
D1   1N4148
D2   1N4148
TR1   BC549
TR2   BC559
TR3   BC549

Miscellaneous
SK1   5 way (180 degree) DIN socket
S1 to S3   Hex switch
S4 to S7   Hex switch
X1   4 MHz crystal
8 pin d.i.l. holder
14 pin d.i.l. holder
16 pin d.i.l. holder (2 off)
18 pin d.i.l. holder

Miscellaneous (continued)
20 pin d.i.l. holder (3 off)
40 pin d.i.l. holder
Case, circuit board, wire, etc.

Additional Components for Mute Switch (Fig.3.21)

Resistor (0.25 watt 5% carbon film)
Ra   4k7

Miscellaneous
Sa   SPST switch

MIDI Processing

A lot of MIDI add-ons are concerned with the processing of MIDI data. Units of this type provide a wide range of processing types, including such things as filtering certain types of MIDI message, harmonising (changing note on and note off messages to alter their pitch value), and channelising (changing the channel number of messages which are on a certain channel or channels). Some types of MIDI processing are beyond the scope of simple add-ons, and require micro based hardware running some quite sophisticated software. However, there are some useful functions that can be provided by relatively simple and non-micro based hardware, and one of these is channelising.

Probably the most common use of channelising is in conjunction with an instrument which only provides operation on channel 1, to effectively enable it to operate on other channels. Few (if any) current instruments have this restriction. It was quite a common constraint in the early days of MIDI, and is one that afflicted some instruments until quite recently. Consequently, there are quite a few MIDI equipped instruments in circulation that give a "Ford" choice, operation on any channel, provided it is channel 1!

There are two basic types of channelising. The first is for use with an instrument which can only receive on MIDI channel 1. The channeliser must let most MIDI messages pass without processing them in any way. It must only process messages that are on a user specified channel, and it must change those messages to MIDI channel 1. This enables the instrument to operate on any MIDI channel.

Ideally the unit would block messages on MIDI channel 1, or move them onto the user specified input channel. This would then leave channel 1 free for use. The simple channeliser described here allows messages on channel 1 to pass unaltered. Therefore, channel 1 must be left unused, and if there are two or more instruments in the system that only give operation on channel 1, these must each be preceded by a channeliser. There are actually other permutations. Consider the situation where there are two instruments in the system operating on channel 1, with one unit fed direct from the controller and the other fed via the channeliser. Messages on channel 1 would be fed to both instruments while those shifted to channel 1 via the channeliser would only be fed to that particular instrument. This would give the option of directing messages to one or both instruments — something that is not normally possible except by doubling-up messages on two channels.

For effective operation with an instrument that can only transmit on channel 1, a channeliser simply needs to convert any MIDI channel messages to a user specified channel. It does not need to filter out messages that are not on channel 1, or process them differently, since it will never be fed with messages on any other channel. This enables a somewhat more simple circuit to be used. This channeliser normally only alters messages on a specified channel. However, as described later in this chapter, if it is only required for use at the output of a single channel MIDI instrument, it can be used in somewhat simplified form.

System Operation

The block diagram of Figure 3.22 shows the basic arrangement used in the channeliser, albeit in rather over-simplified form. On the output side of the circuit there is the transmitter section of the UART, together with the clock oscillator and an invert/driver at the output. Unlike the previous projects, this one also utilizes the receiver section of the UART. This is preceded by an opto-isolator circuit which converts the incoming signal into a form that is suitable for connection to the

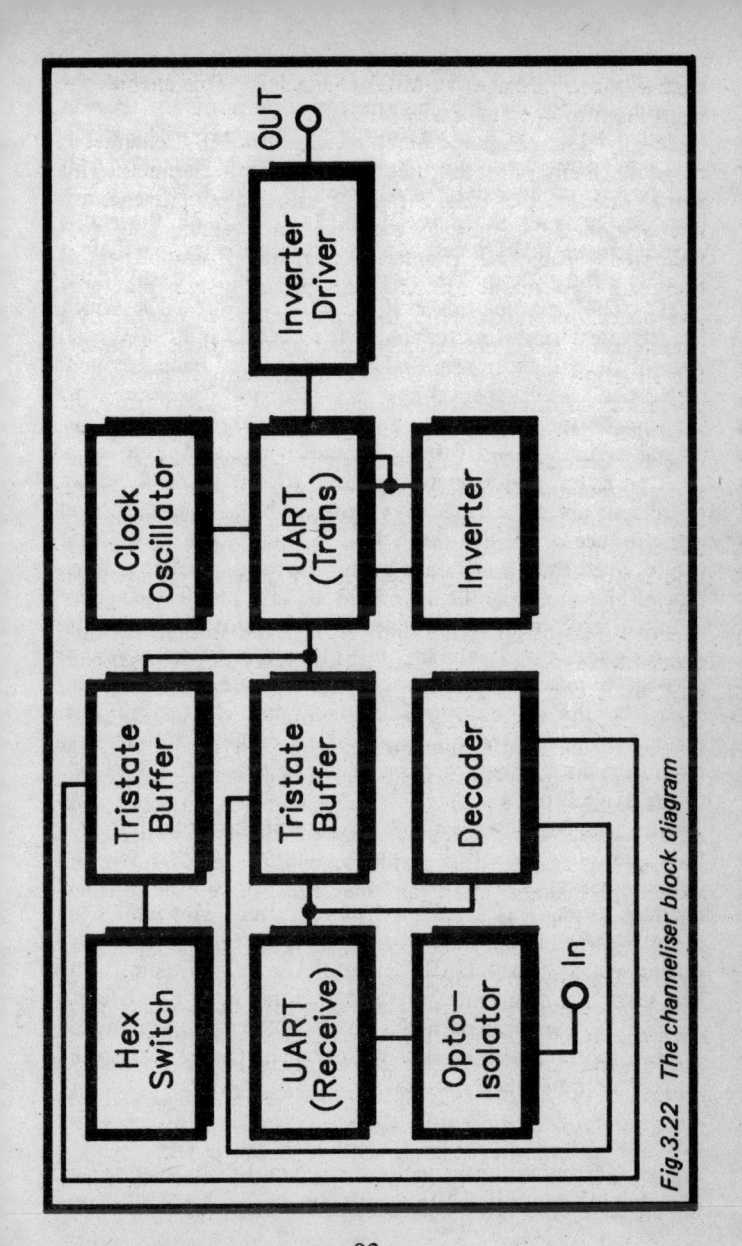

Fig.3.22 The channeliser block diagram

serial input of the UART. It uses the same clock signal and control register as the transmitter section of the device.

The UART has a status output ("data received") which goes high when a complete byte of data has been received. It also has an input ("data received reset") which is taken low in order to reset this status flag. In this circuit the status output drives its own reset input via an inverter, so that as soon as a full byte of data has been received the status flag is reset. The point of this is to generate a short pulse which triggers the transmitter section of the UART, and causes each byte of data to be transmitted as soon as it has been decoded.

Normally each received byte is passed from the receiver to the transmitter via a tristate buffer, and is immediately transmitted in unmodified form. A decoder circuit monitors the received bytes, and activates a second tristate buffer if certain conditions are met. The most significant bit is monitored, and will produce a positive result if it is high (which means that the received byte is a header byte). However, the other three bits of the most significant nibble are also monitored, and a negative result will be produced if all three of these plus the most significant bit are high. This occurs when a system message is received. It is important that these are not processed by the unit as they do not contain a channel number. Instead, the least significant nibble contains a code that identifies the message type, and altering this would change the message type.

The least significant nibble is monitored to see if it contains the user specified channel number. As the decoder circuit detects a channel message byte on the appropriate channel it activates a second tristate buffer. This allows the most significant nibble to pass straight through to the transmitter section of the UART, as before. A hex switch provides the least significant nibble, and this replaces the channel number in the received message. You therefore dial up on the hex switch the channel number you want to be substituted on processed header bytes.

The Circuit
Figure 3.23 shows the main circuit diagram for the channeliser. This has obvious similarities with the previous UART based

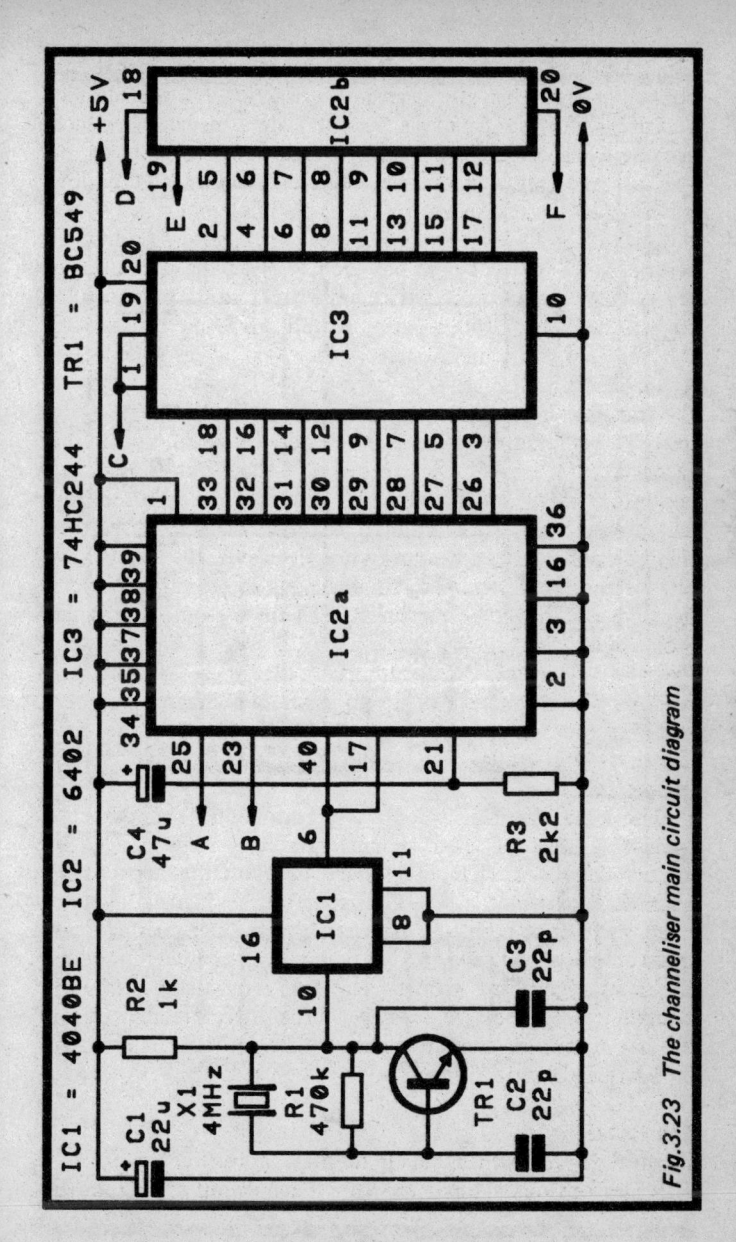

Fig.3.23 The channeliser main circuit diagram

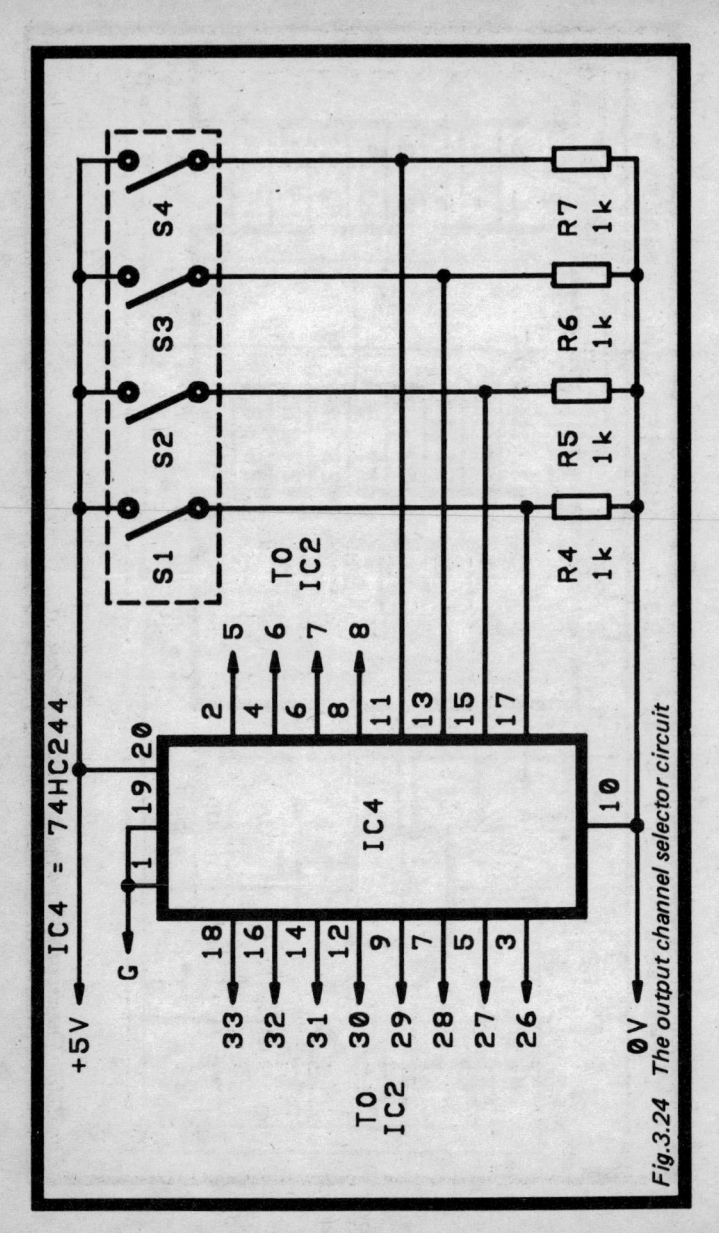

Fig.3.24 The output channel selector circuit

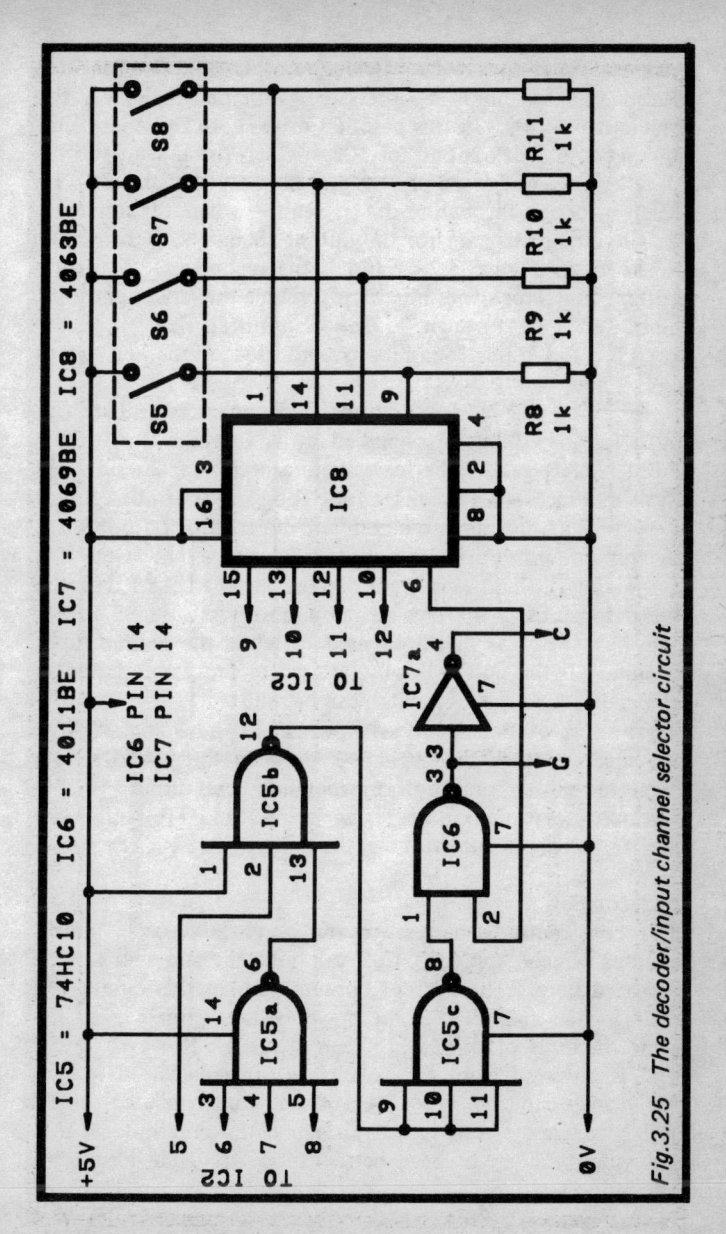

Fig.3.25 The decoder/input channel selector circuit

projects, and the clock, divider, and UART (transmitter) sections of this circuit are the same as their equivalents in the previous projects. In this case IC1 drives both the transmitter and receiver clock inputs of IC2. IC3 is the first of the tristate buffers, and it simply couples the output of the UART's receiver section through to the transmitter when it is activated.

The circuit diagram for the output channel selector circuit is shown in Figure 3.24. IC4 is the second of the tristate buffers, and four of its bits simply couple the most significant nibble straight through to the transmitter section of the UART. The other four bits are provided by the hex switch circuit based on S1 to S4.

Refer to Figure 3.25 for the decoder circuit. The four most significant bits are handled by IC5, which is a triple 3 input NAND gate. The least significant nibble is monitored by IC8, which is a 4 bit magnitude comparator. In this circuit it is used in the mode where it provides a high output when the bit patterns on its two sets of inputs match. Here it is comparing the least significant nibble from the UART with the bit pattern from the hex switch circuit based on S5 to S8. The hex switch is therefore used to select the desired input channel. If the unit will only be used to process an input on channel 1, then the hex switch can be omitted.

The rest of the circuit is shown in Figure 3.26. IC7b is the inverter stage that connects between the UART's data received output and the corresponding reset input. IC9 is the opto-isolator input stage, while TR2 and TR3 act as the inverter/driver at the output of the unit.

Construction

The constructional notes for the previous projects in this chapter largely apply to this unit as well, and will not be repeated here. The current consumption of this circuit is a little higher than that of the previous three projects, but the same methods of powering it can be used. However, if the unit is powered from a 9 volt battery it would need to be a fairly high capacity type. A 100 milliamp regulator should still be adequate, but it will probably run quite hot. A 500 milliamp type might be a better choice for good long term reliability.

Fig.3.26 The inverter, input, and output circuits

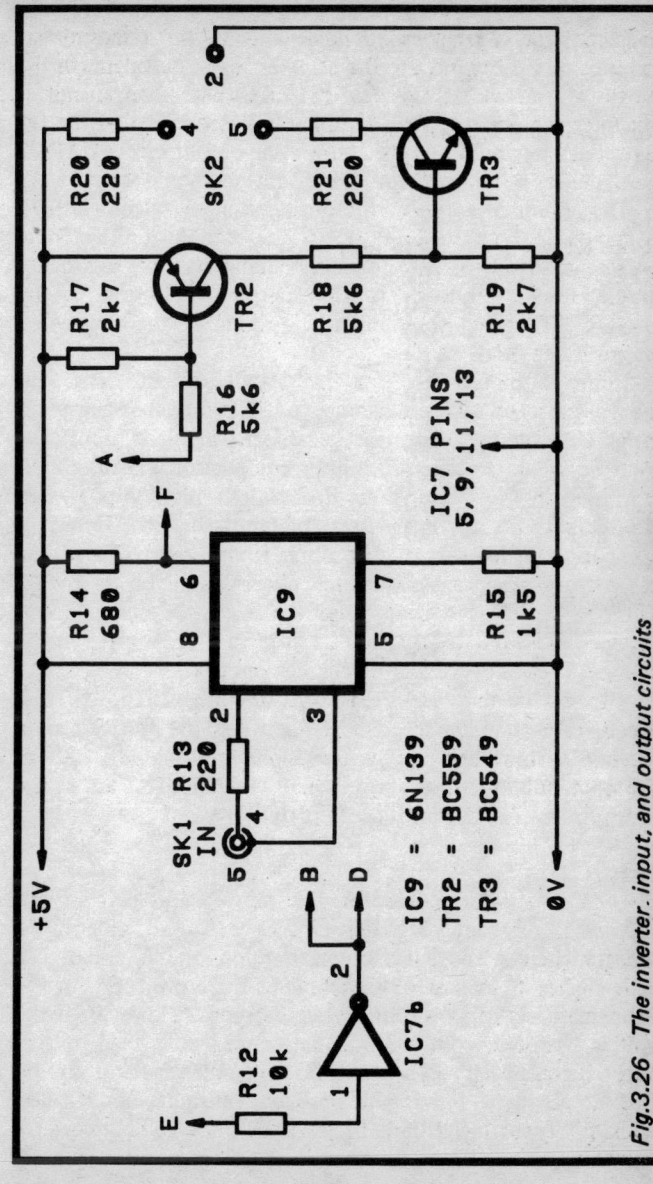

Components for MIDI Channeliser
(Figs 3.23, 3.24, 3.25 & 3.26)

Resistors (all 0.25 watt 5% carbon film)
R1	470k
R2	1k
R3	2k2
R4 to R11	1k (8 off)
R12	10k
R13	220
R14	680
R15	1k5
R16	5k6
R17	2k7
R18	5k6
R19	2k7
R20	220
R21	220

Capacitors
C1	22μ 16V elect
C2	22p ceramic
C3	22p ceramic
C4	47μ 10V elect

Semiconductors
IC1	4040BE or 74HC4040
IC2	6402
IC3	74HC244
IC4	74HC244
IC5	74HC10
IC6	4011BE
IC7	4069BE
IC8	4063BE
IC9	6N139
TR1	BC549
TR2	BC559
TR3	BC549

Miscellaneous

S1 to S4	Hex switch
S5 to S8	Hex switch
SK1	5 way (180 degree) DIN socket
SK2	5 way (180 degree) DIN socket
X1	4 MHz crystal

8 pin d.i.l. holder
14 pin d.i.l. holder (3 off)
16 pin d.i.l. holder (2 off)
20 pin d.i.l. holder (2 off)
40 pin d.i.l. holder
Case, circuit board, wire, etc.

In Use

If the unit is to be used to process incoming signals, its IN socket is fed with the input signal, and its OUT socket connects to the IN socket of the instrument. If it must process outgoing data, its IN socket is fed from the OUT socket of the instrument, and its OUT socket is coupled to the IN socket of the second instrument or other equipment that must be fed with the processed signal. Of course, it is perfectly acceptable to process both incoming and outgoing messages, but two units will be required to do this. Once everything is wired up correctly it is then just a matter of trying out the unit to ensure that it provides the required channel shifting action. Remember that it will only change messages on the channel selected using S5 to S8 to the one selected using S1 to S4. All messages on the wrong input channel pass through the unit without undergoing any changes.

Simplification

If the unit will only be used at the output of an instrument to shift its output to a different channel, a somewhat simplified circuit can be used. Pin 3 of IC7 and point "G" (Fig.3.24) should be fed from pin 12 of IC5. IC6, IC8, S5 to S8, and R8 to R11 can then be omitted. The action of the unit is then to convert all MIDI channel messages to the channel set on S1 to S4, no matter what input channel they are on.

MIDI Analyser

When putting together complex MIDI systems and trying to get everything set up correctly it can sometimes be difficult to track down malfunctions. Most of these faults are not actually faults at all — it is just that something in the system is not set to the right operating mode, or sub-mode of an operating mode. With modern MIDI equipment and software there are usually a large number of options to choose from, and it is very easy to overlook something when setting-up a system. It can be quite time consuming (and frustrating) to track down these errors. Is it the transmitting device or the receiving one which is set to the wrong mode, or is there a genuine fault in the system?

This unit helps with the tracking down of problems in MIDI systems by indicating what type of MIDI message or messages a MIDI source is producing. With channel messages it shows the type of message (note on, pitch wheel, etc.) plus its MIDI channel number. For system messages it shows that the message type is indeed of the system variety, and exactly what kind of system message it is (start, continue, etc.). If the MIDI source is sending data on the wrong channel or something of this nature, this analyser should quickly identify the problem. The message type and channel number are indicated on a twenty-four LED display.

System Operation

The block diagram of Figure 3.27 shows the general arrangement used in this project. It is based on a UART which has a clock oscillator circuit to set the correct baud rate, and an opto-isolator circuit to convert incoming signals into a form that the UART can read. In this application only the receiver section of the UART is utilized.

The UART gives a series of eight bit codes on its parallel output, and the rest of the circuit must filter out the header bytes from the data bytes, and provide information on the header byte. This processing is done by two decoder circuits. The first of these detects message bytes, and indicates the type of message by setting one of eight outputs high. It is easy for the unit to differentiate between header bytes and data types as the most significant bit is always high on the former and

Fig.3.27 The MIDI analyser block diagram

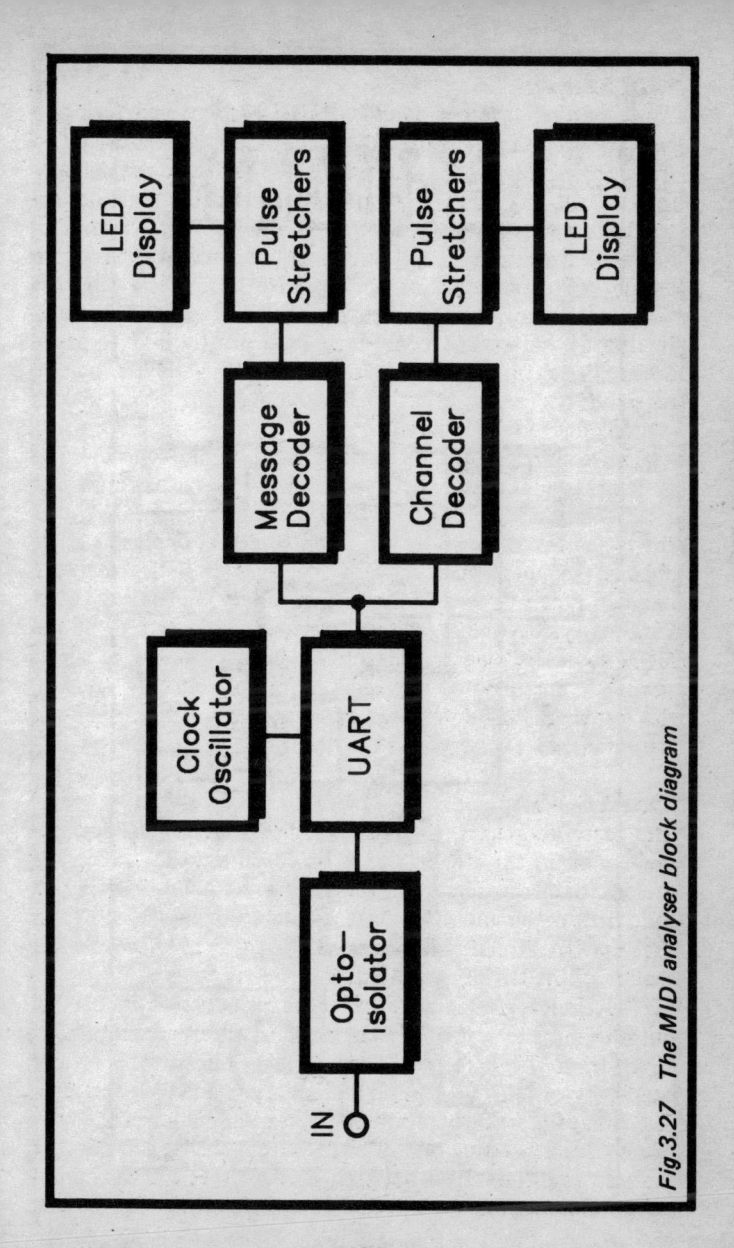

Fig.3.28 The MIDI analyser input circuit

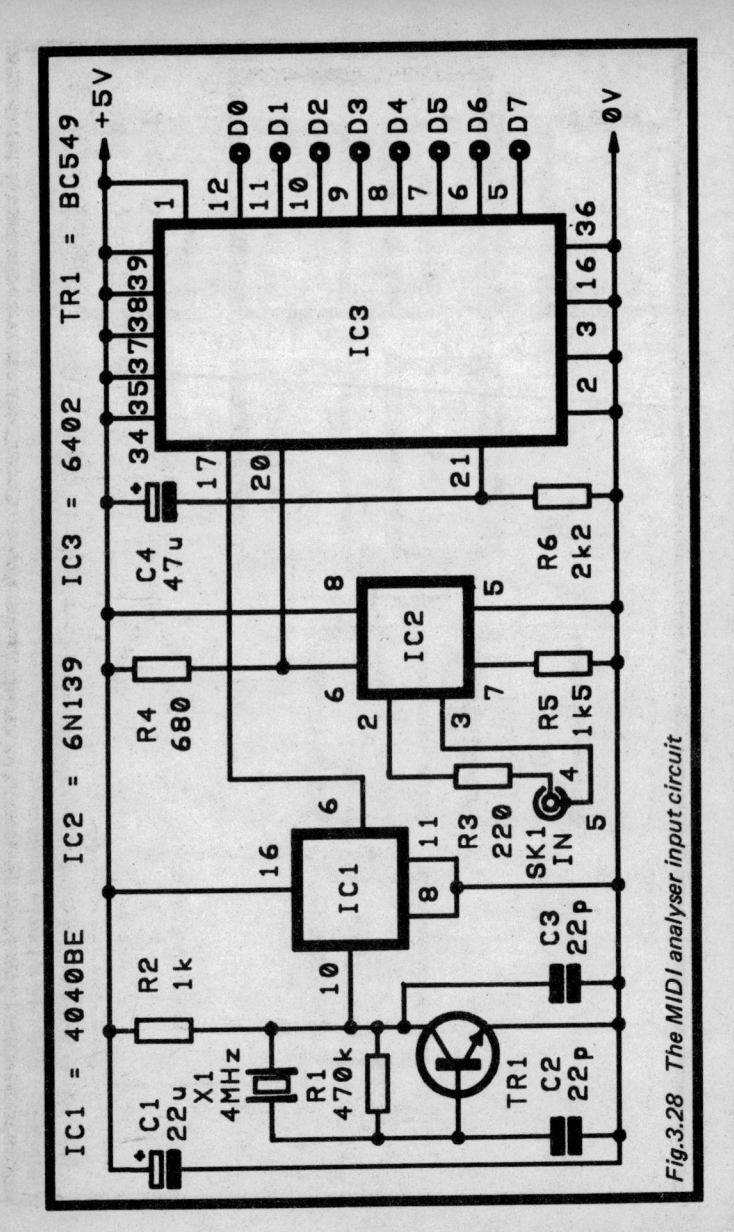

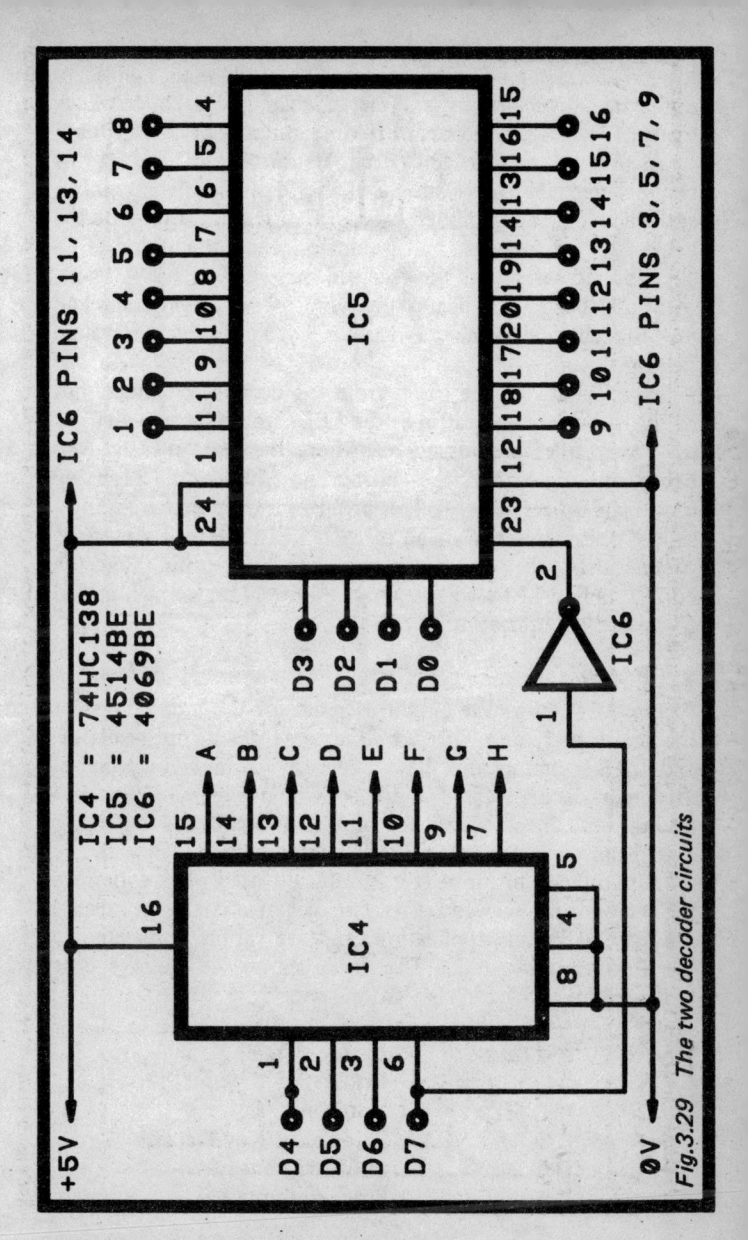

Fig.3.29 The two decoder circuits

low on the latter. The other three bits of the most significant nibble are processed by a three to eight line decoder. Each message type activates a different output of this circuit.

A four to sixteen line decoder forms the basis of the channel decoder. This monitors the least significant nibble, and each channel number causes a different output of the device to be activated. Of course, when the other MIDI decoder indicates that the received message is a system type, then this decoder indicates the type of system message and not the channel number (which is not applicable to system messages).

Directly driving the LEDs from the decoder circuits is not likely to be very successful as the LEDs would be activated for only very brief periods. Even where the same message was being sent repeatedly, the appropriate LEDs would probably only light up very dimly. This problem is overcome by using a simple pulse stretcher ahead of each LED. Even if a message occurs just once, the pulse stretchers will ensure that the corresponding LEDs will be activated for a long enough period to give a clear indication.

The Circuit

Figure 3.28 shows the circuit diagram for the input stages of the MIDI analyser. This uses the same clock, opto-isolator, and UART circuits that have been used in previous projects. The two decoder circuits are shown in Figure 3.29. IC4 decodes the most significant nibble, and this is a 74HC138 3 to 8 line decoder. Its positive enable input is fed with the most significant bit, so that it is only activated when a message header byte is received. The two negative enable inputs are not needed in this application, and are simply connected to the 0 volt supply rail. This table shows the message type indicated by each of IC4's eight outputs.

IC4 Output	*Message Type*
A	Note Off
B	Note On
C	Polyphonic Key Pressure
D	Control Change
E	Program Change

IC4 Output	Message Type
F	Channel Pressure
G	Pitch Wheel Change
H	System Message

The second decoder is based on IC5, which is a 4514BE 4 to 16 line decoder. This decodes the least significant nibble. It has an "inhibit" input at pin 23, and this is driven from the most significant bit via an inverter (IC6). Consequently, IC5 is deactivated when data bytes are present on the output of IC3. The numbers marked on the outputs of IC5 show the MIDI channel* number that activates each one. When the message is a system type, the outputs indicate the kind of system message that has been received, as detailed in this table. Note that some of the available codes are not yet assigned, and have therefore been omitted from this table.

IC5 Output	Message Type
1	System Exclusive
3	Song Position Pointer
4	Song Select
7	Tune Request
8	End Of System Exclusive
9	Clock
11	Start
12	Continue
13	Stop
15	Active Sensing
16	System Reset

There is an important difference between IC4 and IC5 in that IC4's outputs are normally high and go low when activated, whereas IC5's outputs are normally low and go high when activated. Consequently they require slightly different pulse stretcher/LED driver circuits. IC4 requires the stretcher circuit shown in Figure 3.30. There are four pulse stretchers here, with one based on each of IC7's gates. These are NAND gates, but in this circuit they are wired to operate as simple inverters. The diode, resistor, and capacitor at the input of each circuit, aided by the very high input impedance of

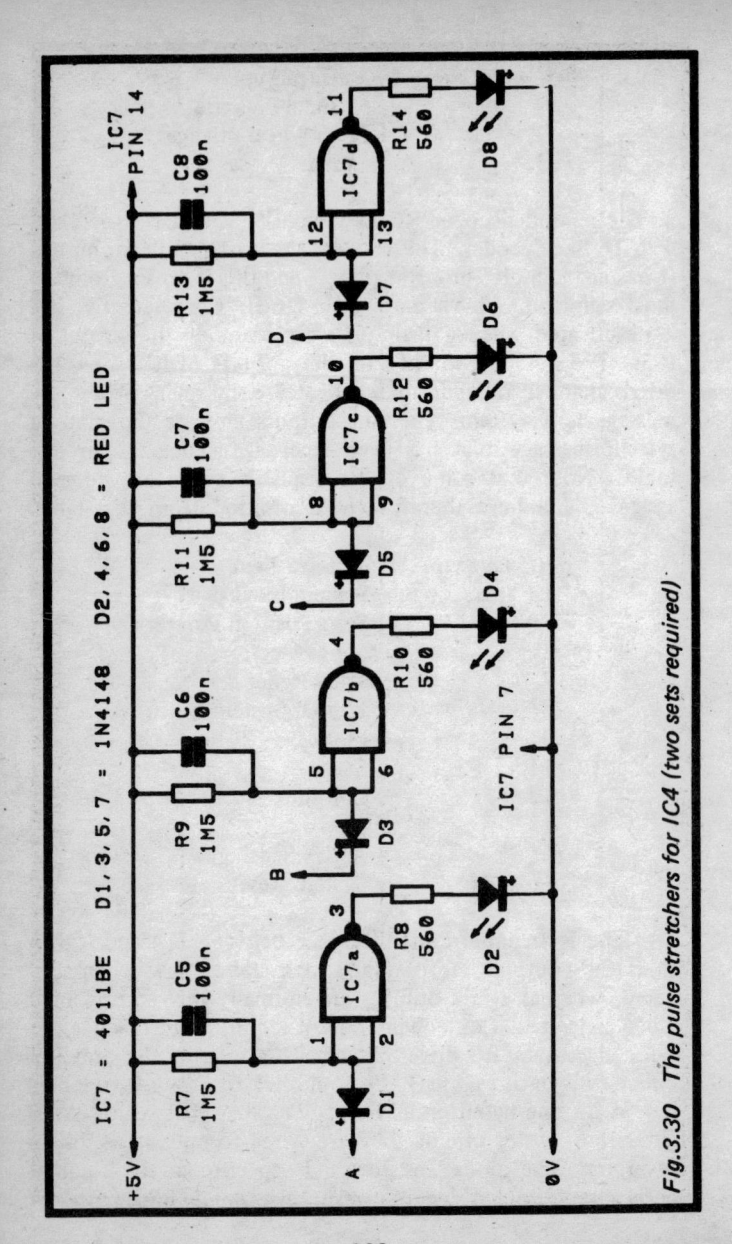

Fig.3.30 The pulse stretchers for IC4 (two sets required).

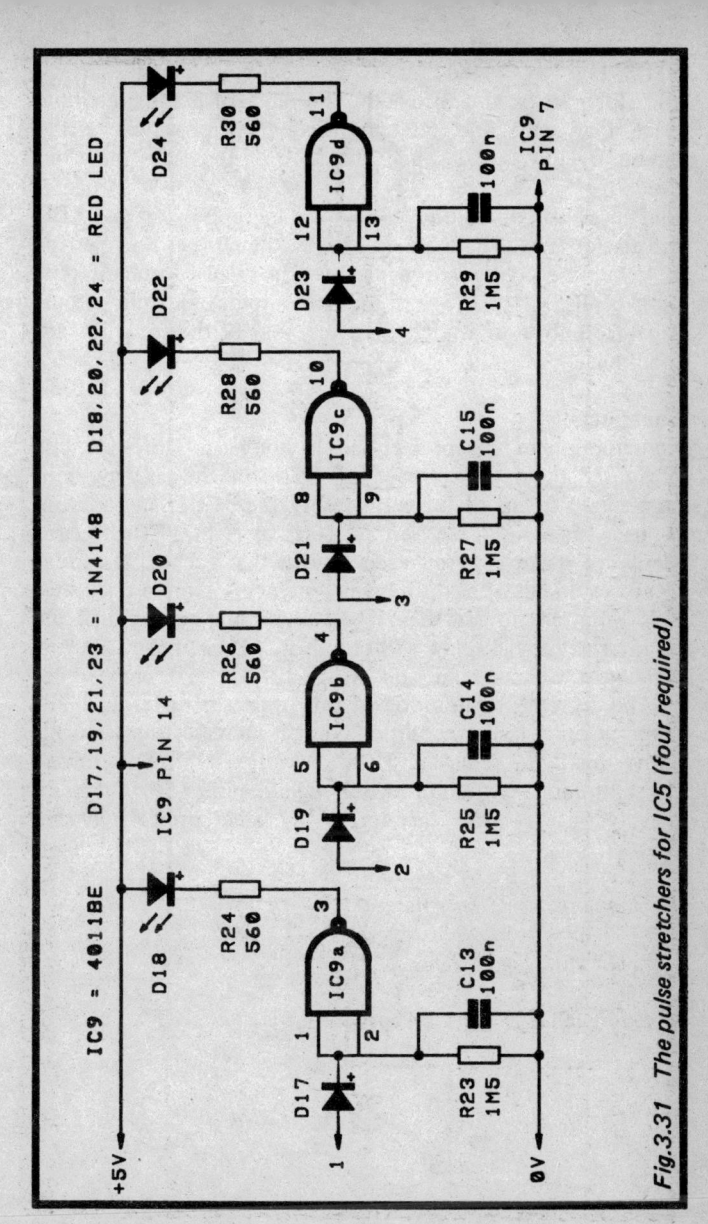

Fig.3.31 The pulse stretchers for IC5 (four required)

CMOS logic integrated circuits, give these circuits a fast attack and slow decay, thus providing the required pulse stretching. Note that only four stretcher/drivers are provided by the circuit of Figure 3.29, and that two of these are therefore needed.

The circuit of Figure 3.31 shows the pulse stretcher/LED drivers for IC5. These are essentially the same as the ones for IC4, but the configuration has been inverted to suit the outputs of IC5. As IC5 has sixteen outputs, and this circuit provides only four stretcher/drivers, four of these circuits are required.

Construction

Construction of this project should not be too difficult, but obviously the large number of LEDs in the display does complicate things slightly. Probably the best arrangement is to have them in three vertical rows of eight LEDs. Leave plenty of space between each row so that each LED can be clearly marked with the channel number it represents, etc. It is important to add these labels as it would be very difficult to interpret the display without them. Even if they are not very neat they should still prove to be very worthwhile.

The current consumption of this project is under 20 milli-amps under standby conditions, but it increases substantially above this figure if several LEDs are switched on. The current consumption is still quite low enough to permit the unit to be powered in the same manner as the other projects in this chapter.

Components for MIDI Analyser

(Main Circuit, Figs 3.28 & 3.29)

Resistors (all 0.25 watt 5% carbon film)
R1	470k
R2	1k
R3	220
R4	680
R5	1k5
R6	2k2

Capacitors
C1	22µ 16V elect
C2	22p ceramic
C3	22p ceramic
C4	47µ 10V elect

Semiconductors
IC1	4040BE or 74HC4040
IC2	6N139
IC3	6402
IC4	74HC138
IC5	4514BE
IC6	4069BE
TR1	BC549

Miscellaneous
SK1	5 way (180 degree) DIN socket
X1	4 MHz crystal

Case, circuit board, wire etc.

(Display, Fig.3.30, also component values apply to Fig.3.31)

Resistors (all 0.25 watt 5% carbon film)
R7	1M5
R8	560
R9	1M5
R10	560
R11	1M5
R12	560
R13	1M5
R14	560

Capacitors
C5	100n polyester
C6	100n polyester
C7	100n polyester
C8	100n polyester

Semiconductors
IC7	4011BE

Semiconductors (continued)
D1   1N4148
D2   Red LED
D3   1N4148
D4   Red LED
D5   1N4148
D6   Red LED
D7   1N4148
D8   Red LED

Miscellaneous
LED panel holder (4 off)
14 pin d.i.l. holder
Wire, solder, etc.

Note that six sets of display components are needed in order to provide a full twenty-four LED display.

Finally
The projects featured in this book are all tried and tested designs, but they also provide a useful selection of basic building blocks which experienced readers can use as the basis of their own designs. Something that is well worth pursuing is MIDI processing. Apart from channelising, there are other possible applications for simple processors, including MIDI filters and harmonisers. There is plenty of scope for experimentation, and MIDI enables the imaginative user to do practically anything he or she wishes to.

Semiconductor pinout details are shown in Figure 3.32.

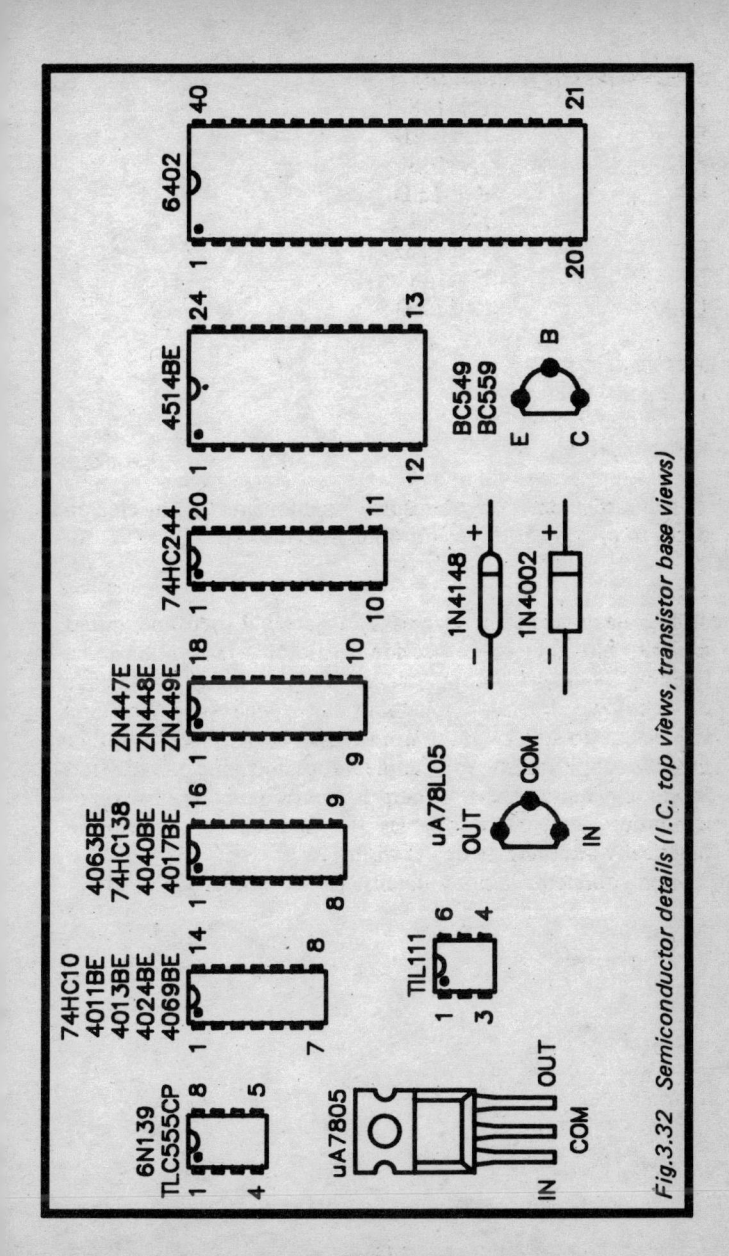

Fig.3.32 Semiconductor details (I.C. top views, transistor base views)

Notes

Notes

Notes

Notes

Please note following is a list of other titles that are available in our range of Radio, Electronics and Computer Books.

These should be available from all good Booksellers, Radio Component Dealers and Mail Order Companies.

However, should you experience difficulty in obtaining any title in your area, then please write directly to the publisher enclosing payment to cover the cost of the book plus adequate postage.

If you would like a complete catalogue of our entire range of Radio, Electronics and Computer Books then please send a Stamped Addressed Envelope to:

BERNARD BABANI (publishing) LTD
THE GRAMPIANS
SHEPHERDS BUSH ROAD
LONDON W6 7NF
ENGLAND

Code	Title	Price
160	Coil Design and Construction Manual	£2.50
205	Hi-Fi Loudspeaker Enclosures	£2.95
208	Practical Stereo & Quadrophony Handbook	£0.75
214	Audio Enthusiast's Handbook	£0.85
219	Solid State Novelty Projects	£0.85
220	Build Your Own Solid State Hi-Fi and Audio Accessories	£0.85
222	Solid State Short Wave Receivers for Beginners	£2.95
225	A Practical Introduction to Digital ICs	£2.50
226	How to Build Advanced Short Wave Receivers	£2.95
227	Beginners Guide to Building Electronic Projects	£1.95
228	Essential Theory for the Electronics Hobbyist	£2.50
BP2	Handbook of Radio, TV, Industrial and Transmitting Tube and Valve Equivalents	£0.60
BP6	Engineer's & Machinist's Reference Tables	£1.25
BP7	Radio & Electronic Colour Codes Data Chart	£0.95
BP27	Chart of Radio, Electronic, Semiconductor and Logic Symbols	£0.95
BP28	Resistor Selection Handbook	£0.60
BP29	Major Solid State Audio Hi-Fi Construction Projects	£0.85
BP33	Electronic Calculator Users Handbook	£1.50
BP36	50 Circuits Using Germanium Silicon and Zener Diodes	£1.50
BP37	50 Projects Using Relays, SCRs and TRIACs	£2.95
BP39	50 (FET) Field Effect Transistor Projects	£2.95
BP42	50 Simple LED Circuits	£1.95
BP44	IC 555 Projects	£2.95
BP45	Projects in Opto-Electronics	£1.95
BP48	Electronic Projects for Beginners	£1.95
BP49	Popular Electronic Projects	£2.50
BP53	Practical Electronics Calculations and Formulae	£3.95
BP54	Your Electronic Calculator & Your Money	£1.35
BP56	Electronic Security Devices	£2.50
BP58	50 Circuits Using 7400 Series IC's	£2.50
BP62	The Simple Electronic Circuit & Components (Elements of Electronics – Book 1)	£3.50
BP63	Alternating Current Theory (Elements of Electronics – Book 2)	£3.50
BP64	Semiconductor Technology (Elements of Electronics – Book 3)	£3.50
BP66	Beginners Guide to Microprocessors and Computing	£1.95
BP68	Choosing and Using Your Hi-Fi	£1.65
BP69	Electronic Games	£1.75
BP70	Transistor Radio Fault-finding Chart	£0.95
BP72	A Microprocessor Primer	£1.75
BP74	Electronic Music Projects	£2.50
BP76	Power Supply Projects	£2.50
BP77	Microprocessing Systems and Circuits (Elements of Electronics – Book 4)	£2.95
BP78	Practical Computer Experiments	£1.75
BP80	Popular Electronic Circuits – Book 1	£2.95
BP84	Digital IC Projects	£1.95
BP85	International Transistor Equivalents Guide	£3.50
BP86	An Introduction to BASIC Programming Techniques	£1.95
BP87	50 Simple LED Circuits – Book 2	£1.35
BP88	How to Use Op-Amps	£2.95
BP89	Communication (Elements of Electronics – Book 5)	£2.95
BP90	Audio Projects	£2.50
BP91	An Introduction to Radio DXing	£1.95
BP92	Electronics Simplified – Crystal Set Construction	£1.75
BP93	Electronic Timer Projects	£1.95
BP94	Electronic Projects for Cars and Boats	£1.95
BP95	Model Railway Projects	£1.95
BP97	IC Projects for Beginners	£1.95
BP98	Popular Electronic Circuits – Book 2	£2.25
BP99	Mini-matrix Board Projects	£2.50
BP101	How to Identify Unmarked ICs	£0.95
BP103	Multi-circuit Board Projects	£2.25
BP104	Electronic Science Projects	£2.95
BP105	Aerial Projects	£1.95
BP106	Modern Op-amp Projects	£1.95
BP107	30 Solderless Breadboard Projects – Book 1	£2.25
BP108	International Diode Equivalents Guide	£2.25
BP109	The Art of Programming the 1K ZX81	£1.95
BP110	How to Get Your Electronic Projects Working	£2.50
BP111	Audio (Elements of Electronics – Book 6)	£3.50
BP112	A Z-80 Workshop Manual	£3.50
BP113	30 Solderless Breadboard Projects – Book 2	£2.25
BP114	The Art of Programming the 16K ZX81	£2.50
BP115	The Pre-computer Book	£1.95
BP117	Practical Electronic Building Blocks – Book 1	£1.95
BP118	Practical Electronic Building Blocks – Book 2	£1.95
BP119	The Art of Programming the ZX Spectrum	£2.50
BP120	Audio Amplifier Fault-finding Chart	£0.95
BP121	How to Design and Make Your Own PCB's	£1.95
BP122	Audio Amplifier Construction	£2.25
BP123	A Practical Introduction to Microprocessors	£2.50
BP124	Easy Add-on Projects for Spectrum, ZX81 & Ace	£2.75
BP125	25 Simple Amateur Band Aerials	£1.95
BP126	BASIC & PASCAL in Parallel	£1.50
BP127	How to Design Electronic Projects	£2.25
BP128	20 Programs for the ZX Spectrum and 16K ZX81	£1.95
BP129	An Introduction to Programming the ORIC-1	£1.95
BP130	Micro Interfacing Circuits – Book 1	£2.25
BP131	Micro Interfacing Circuits – Book 2	£2.75

BP132	25 Simple Shortwave Broadcast Band Aerials	£1.95
BP133	An Introduction to Programming the Dragon 32	£1.95
BP135	Secrets of the Commodore 64	£1.95
BP136	25 Simple Indoor and Window Aerials	£1.75
BP137	BASIC & FORTRAN in Parallel	£1.95
BP138	BASIC & FORTH in Parallel	£1.95
BP139	An Introduction to Programming the BBC Model B Micro	£1.95
BP140	Digital IC Equivalents & Pin Connections	£5.95
BP141	Linear IC Equivalents & Pin Connections	£5.95
BP142	An Introduction to Programming the Acorn Electron	£1.95
BP143	An Introduction to Programming the Atari 600/800XL	£1.95
BP144	Further Practical Electronics Calculations and Formulae	£4.95
BP145	25 Simple Tropical and MW Band Aerials	£1.75
BP146	The Pre-BASIC Book	£2.95
BP147	An Introduction to 6502 Machine Code	£2.50
BP148	Computer Terminology Explained	£1.95
BP149	A Concise Introduction to the Language of BBC BASIC	£1.95
BP152	An Introduction to Z80 Machine Code	£2.75
BP153	An Introduction to Programming the Amstrad CPC464 and 664	£2.50
BP154	An Introduction to MSX BASIC	£2.50
BP156	An Introduction to QL Machine Code	£2.50
BP157	How to Write ZX Spectrum and Spectrum+ Games Programs	£2.50
BP158	An Introduction to Programming the Commodore 16 and Plus 4	£2.50
BP159	How to write Amstrad CPC 464 Games Programs	£2.50
BP161	Into the QL Archive	£2.50
BP162	Counting on QL Abacus	£2.50
BP169	How to Get Your Computer Programs Running	£2.50
BP170	An Introduction to Computer Peripherals	£2.50
BP171	Easy Add-on Projects for Amstrad CPC 464, 664, 6128 and MSX Computers	£3.50
BP173	Computer Music Projects	£2.95
BP174	More Advanced Electronic Music Projects	£2.95
BP175	How to Write Word Game Programs for the Amstrad CPC 464, 664 and 6128	£2.95
BP176	A TV-DXers Handbook	£5.95
BP177	An Introduction to Computer Communications	£2.95
BP179	Electronic Circuits for the Computer Control of Robots	£2.95
BP180	Electronic Circuits for the Computer Control of Model Railways	£2.95
BP181	Getting the Most from Your Printer	£2.95
BP182	MIDI Projects	£2.95
BP183	An Introduction to CP/M	£2.95
BP184	An Introduction to 68000 Assembly Language	£2.95
BP185	Electronic Synthesiser Construction	£2.95
BP186	Walkie-Talkie Projects	£2.95
BP187	A Practical Reference Guide to Word Processing on the Amstrad PCW8256 & PCW8512	£5.95
BP188	Getting Started with BASIC and LOGO on the Amstrad PCWs	£5.95
BP189	Using Your Amstrad CPC Disc Drives	£2.95
BP190	More Advanced Electronic Security Projects	£2.95
BP191	Simple Applications of the Amstrad CPCs for Writers	£2.95
BP192	More Advanced Power Supply Projects	£2.95
BP193	LOGO for Beginners	£2.95
BP194	Modern Opto Device Projects	£2.95
BP195	An Introduction to Satellite Television	£5.95
BP196	BASIC & LOGO in Parallel	£2.95
BP197	An Introduction to the Amstrad PC's	£5.95
BP198	An Introduction to Antenna Theory	£2.95
BP199	An Introduction to BASIC-2 on the Amstrad PC's	£5.95
BP230	An Introduction to GEM	£5.95
BP232	A Concise Introduction to MS-DOS	£2.95
BP233	Electronic Hobbyists Handbook	£4.95
BP234	Transistor Selector Guide	£4.95
BP235	Power Selector Guide	£4.95
BP236	Digital IC Selector Guide-Part 1	£4.95
BP237	Digital IC Selector Guide-Part 2	£4.95
BP238	Linear IC Selector Guide	£4.95
BP239	Getting the Most from Your Multimeter	£2.95
BP240	Remote Control Handbook	£3.95
BP241	An Introduction to 8086 Machine Code	£5.95
BP242	An Introduction to Computer Aided Drawing	£2.95
BP243	BBC BASIC86 on the Amstrad PC's and IBM Compatibles -- Book 1: Language	£3.95
BP244	BBC BASIC86 on the Amstrad PC's and IBM Compatibles -- Book 2: Graphics & Disc Files	£3.95
BP245	Digital Audio Projects	£2.95
BP246	Musical Applications of the Atari ST's	£4.95
BP247	More Advanced MIDI Projects	£2.95
BP248	Test Equipment Construction	£2.95
BP249	More Advanced Test Equipment Construction	£2.95
BP250	Programming in FORTRAN 77	£4.95
BP251	Computer Hobbyists Handbook	£5.95
BP252	An Introduction to C	£2.95
BP253	Ultra High Power Amplifier Construction	£3.95
BP254	From Atoms to Amperes	£2.95
BP255	International Radio Stations Guide	£4.95
BP256	An Introduction to Loudspeakers and Enclosure Design	£2.95
BP257	An Introduction to Amateur Radio	£2.95
BP258	Learning to Program in C	£4.95